THE CASE FOR LOVE

Also by A. K. Benjamin

Let Me Not Be Mad

A. K. BENJAMIN

THE CASE FOR LOVE

THE BODLEY HEAD
LONDON

1 3 5 7 9 10 8 6 4 2

The Bodley Head, an imprint of Vintage, is part of the
Penguin Random House group of companies whose addresses
can be found at global.penguinrandomhouse.com.

Copyright © A K Benjamin 2021

A K Benjamin has asserted his right to be identified as the author of this
Work in accordance with the Copyright, Designs and Patents Act 1988

Image on page 50: 'Over the Town' by Marc Chagall © ADAGP,
Paris and DACS, London 2021.
Image on page 145: 'Blue Angel, Before Leaving. Homage to
Alban Berg Violin Concerto' by © Kazuya Akimoto.

First published in the UK by The Bodley Head in 2021

www.vintage-books.co.uk

A CIP catalogue record for this book is available from the British Library

Hardback ISBN 9781847926197

Printed and bound in Great Britain by Clays Ltd, Elcograf S.p.A.

The authorised representative in the EEA is Penguin Random House Ireland,
Morrison Chambers, 32 Nassau Street, Dublin D02 YH68.

Penguin Random House is committed to a sustainable future for
our business, our readers and our planet. This book is made from
Forest Stewardship Council® certified paper.

For my mother, the first to make the case

Symptoms of disease are nothing but a disguised manifestation of the power of love; and all disease is only love transformed.
Thomas Mann, *The Magic Mountain*

Author's Note

In order to ensure that no real person is identifiable from these pages and that any sensitive material on which I have drawn is protected, I have changed names, physical features, backgrounds, locations, nationalities, ethnicities and the key detail of events.

Prologue

The nurses want to kill her. They may not know it, they certainly wouldn't admit it, but murder is the charge that under white-lipped faces passes through their steely hands. It's how you force something to die, he thinks, treat it as though it was dead already, day after day; they may as well be heaving rolled-up carpet, scouring a kitchen surface.

It meant that nobody had looked at her eyes for weeks but they are marvels: almond-shaped, deep violet, set high, imperiously so. Their brilliance changes continuously under the shifting skies of her forehead (the muscles still work there); sad then proud then rueful. The changes are subtle, easy to miss, easier still to ignore, but half an hour ago, hooked by something – a distant memory? the dreadful prospect before her? he can only imagine – they caught fire and the whole room was threaded through them.

Whatever troubles her – he wants to tell her – will not last. She might use her breath to illustrate this principle, because in its coming and going breath prefigures every experience in the world. Breathe in, he might say, breathe in when it's not breath at all but compressed air; when it's not, strictly speaking, hers.

15.40
VC-CMV, VT 400 mL, f-12, FiO₂ .60, + 8 cmH₂O PEEP
No change.

He imagines the feeling of it; snaking currents over the
floor of the lungs, temperatures blent as inside meets out,
tightness with the rising of the diaphragm, the heart buoyed
on its swell; until a moment of fullness, of weightlessness, and
then another – he imagines the feeling for her because otherwise it will be lost – and still another, the gap before exhaling
stretching out further than it should, to the edge of panic,
waiting for the ventilator to begin again.

18.25
VC-CMV, VT 400 mL, f-12, FiO₂ .60, + 8 cmH₂O PEEP
No change.

What decisions there are, what life there is belongs to the
machine behind her head. The key to life a spectral instruction; from left to right the mode of control and tidal volume,
whether the breath should be triggered by time or by effort,
the respiratory rate (twelve per minute); the fraction of inspired
oxygen, the level of humidity compared to room air; the peak
expiratory end pressure.
Jargon, until it happens to you. She has become technical.

14.15
VC-CMV, VT 400 mL, f-12, FiO₂ .60, + 8 cmH₂O PEEP
Bolus given.
Haemodynamically – no change.

Her food is delivered, without a menu, straight to her
stomach through a tube.

02.25

VC-CMV, VT 400 mL, f-12, FiO$_2$ .60, + 8 cmH$_2$O PEEP
Alarm attended.

Alarms ring day and night, with different tones signalling different problems; when inspiratory or expiratory pressure is beyond limits in either direction, or the volume of lung inflation exceeds or drops below a set level, or the frequency is too high or too low, or there is apnea, or she is accidentally disconnected. Not every alarm needs attending. Experience teaches that things tend to settle themselves down of their own accord, that it's important to avoid unnecessary disturbance. On the other side of experience, habituation; the mundane effort needed to keep *hearing* it, shift after shift after shift. Or to hear it and still respond.

22.15
VC-CMV, VT 450 mL, f-12, FiO$_2$ .60, + 8 cmH$_2$O PEEP
No change.

By this point she has a room of her own, set apart from the rest of Intensive Care. A room for extraordinary cases that take longer than usual to play out. The de facto store for faulty equipment when not in use, it has to be cleared for her. A single bed set in the centre, the space around it accentuated by bare white walls. The ventilator stands where a bedside table would. On either side, more machines and drips with fluids of different colours and concentrations are plumbed into her. He means to bring a picture or two in to give the place some colour. A Chagall perhaps, because she reminds him of Bella the painter's wife, if she'd lived a generation longer: black hair that's still only streaked with silver, sharp Slavic cheekbones around her precious eyes, the spell cast by her balefulness. Behind the bed a large single window looks over the town; cranes, high-rises, mosques, terraces of blackening red-brick back-to-backs; the life she never sees. Instead the tilt of the bed means she's forced to look past the coffin of her body through another large window, behind which there is a chair and a desk where her medical file lies open. In this way she watches the nurses convert her aliveness into ventilator codes and basic observations in the medical notes.

He's the clinical neuropsychologist, part of her team, watching through the glass. There was nothing for him to do professionally. Still he sits there whenever he has a spare moment, day after day, week after week, on Saturdays if he doesn't have the kids, watching the dead still centre of the world, as though the reason for his sitting will come. Sometimes he's not alone, there's another figure watching her too, on the other side of the glass. Bella's partner, Marc, say: an older man who always wears the same dirty jeans and hooded top, sometimes sitting on a stool by her feet like a monk keeping vigil at a casket, for as long as he's allowed.

It's disconcerting when he's watching and Marc's there, as though a member of the audience has climbed on stage for a closer look; as though he sees his own absurdity in action.

08.30
VC-CMV, VT 450 mL, f-12, FiO$_2$ .70, + 8 cmH$_2$O PEEP
Bolus given.
No change.

But there *are* changes, he thinks.

Her axis changed a month ago, for the last time. Once tall she would only be long now, and that length itself was changing. She's shrinking. She's losing weight, less than fifty kilograms judging by the impression she makes on the bed sheet. Lying on her back at an angle of thirty degrees to clear the effluent from her throat and lungs, gravity recruited to help with another job she can no longer do herself. But gravity is not just a helpmeet; limp at rest she becomes a nasty puzzle when lifted with the unguessable torsion and flexing of a held fish. More insidiously, over weeks of lying there, gravity has taken what was once lithe, athletic, shy, tender and neutered it, flattened it like putty right before her eyes.

09.30
VC-CMV, VT 450 mL, f-12, FiO₂ .70, + 8 cmH₂O PEEP
No change.

Really '*Everything changes*', the neuropsychologist wants to write in the notes. Her hair has grown over the long weekend while he took his kids to Broadstairs. Still lustrous, a deep black – the silver cow kiss at the front – like a wimple against the white pillow, only wilder, more exuberant. At least by comparison. Her skin has long since lost any kinship with the 'skin-toned' compression socks that pinch off her legs halfway up her calves. It's grey now, draining the colour of what it touches, like the washed-out lilac bed gown, or the off-white incontinence pads. Still a camouflage of sorts, signalling there's nothing to see here, nothing vital within.
He's not fooled.

16.40
VC-CMV, VT 450 mL, f-12, FiO₂ .70, + 8 cmH₂O PEEP
Alarm attended.
No change.

Dying all over again, hour after hour; it's obvious to anyone who stops what they're doing. Therefore nobody does. So she continues to disappear from view, replaced by symbols, abbreviations, numbers, condemned to live the rest of her life in her head, alone.

17.20
VC-CMV, VT 450 mL, f-12, FiO₂ .70, + 8 cmH₂O PEEP
Alarm attended.
No change.

It's left to him – the useless hero – to save her, long after the fact. She was not his last ever case, but close enough, a template for the thousands of others.

He wants to give her a different kind of life support by imagining what it might have been like to be her, who this 'Bella' was before they killed her, what brought her to this, what future ended with it; life written into the pointless ruins of the medical notes, from the other side of the glass, in a different country, years later.

Love in miniature.

17.40
VC-CMV, VT 450 mL, f-12, FiO$_2$ .70, + 8 cmH$_2$O PEEP
No change.

The ghosts of answers are still lying there, if he looks past the useless body, and the lower half of her face which is permanently screwed into an expression of dull terror. The eyes still display a rare intelligence, a defiance, but also occasionally a girlishness too.

There's Marc, he's useful in this reconstruction too; whoever she was is imprinted in him, the proxy for everything she can no longer feel. Drained of life over the weeks she's been there, each day he came to resemble her more closely, as they say partners do.

There's also a page and a half of 'history' near the start of her medical file – which after only six weeks is more than four inches thick – scribbled down carelessly: work details, family tree, a few likes and dislikes, the only anchor points left. In the first days after her admission these little anchors were mentioned at handovers to keep her in mind. But in the repeating they were further compressed, for the sake of ease, until they disappeared: moved from front matter to index to archive.

He will use them as a starting point: brought up an only child by her father, her profession of actuary, her fondness for tennis, meeting Marc late in life, their retirement to the Albanian alps on the border with Montenegro, the stroke which came the same moment she touched the water, on practically the first day of her new life, the image of her body sinking to the bottom of the swimming pool. He will start with what is known and add it to what he sees for himself, moment by moment, coming and going from the portals of her face.

Start from the outside, with the actual, and then imagine inwards, by increments; shaving off a patch of her hair, sawing through her scalp, pulling apart the dura mater, the arachnoid mater, the pia mater – the different layers that protect but also separate – until he's inside; until he's her.

Bella

If it had been a work decision she would have made it her business to know that Montenegro (where Podgorica was the nearest neurosurgical hospital to their Albanian home) was ninth in the WHO statistics on stroke death, accounting for 29.57 per cent of the nation's total mortalities. But the whole point of going there was to leave that kind of knowledge behind. Instead she got the fate she deserved; someone who lived, according to everyone that knew her, 'in her head', suffering a stroke that didn't kill her, not quite, not yet, but cut her off from everything below the top of her spine, just before body turns to mind.

After a four-hour ambulance ride on roads that hadn't been repaired since the war, the surgical trainee – a little too old to be her grandson, but only just – coiled the bleed in the longest operation of his brief career. A London-bound passenger flight was requisitioned – they had to buy out the whole of first class – but only after Marc had spent several hours on the phone with the insurance company proving, with interjections from the surgeon in broken English, that the haemorrhage had a 'non-sporting' cause and would therefore be covered by her policy: 'Not deep-sea diving, diving in a swimming pool ... How deep? ...' Marc wondered if he was speaking to a call centre. 'Two metres, less, one and a half ... No she didn't hit her head on the bottom ... Yes she can swim ... a mile ... front crawl ... yes I'm sure ... No I can't formally prove it.' He wondered if he was speaking to a person. 'Listen, this will be life insurance unless we get her home immediately ...'

Bella had specialised in predicting catastrophe for insurers for whom futures were merely broad, lifeless categories in a model, until the moment they suddenly arrived, the details sickeningly filled in. She knew when they bought the place the swimming pool would be trouble, just about covered by the terms of 'domestic leisure'. What she couldn't have known was the hours it took for the insurer to arrive at the same conclusion, at a time when hours meant oedema, rising intra-cranial pressure, irreparable cell death. Though the voice on the telephone didn't ask anything directly, Marc could tell it was angling the questions towards mental health, wondering aloud why anyone would *choose* to live on the Albanian border with Montenegro, the graveyard of a decade-long war, where instead of advanced medical technology in premium hospitals there were tree-bound husbands shaking plums from the branches into their wive's bulging skirts, while their children sat in shacks learning English from American reality shows on plasma televisions the size of dinner tables. The insurer had a point; Marc had just gone along with it. Bella's deci-sions, often long-pondered in solitude, were always enough for him. If sanity meant taking different perspectives into account, including the effect of those perspectives on what they saw, then she was sane, relentlessly so, the sanest person he'd ever met. It had driven her mad. It was high on the spreadsheet of reasons she had for leaving that world behind, moving to the middle of nowhere and diving into cool thought-less water at the first available opportunity.

Back in the UK there was still a 200-mile ambulance ride up the M1, and then fifty more miles on A roads to the regional neuroscience centre nearest her home. Cracked concrete, dented steel in the middle of the lanes, broken LED signs that always said 'Accident ahead, 20'; even with the siren on the unending roadworks, the weight of traffic, other near catastrophes, pushed the journey time to six hours. So much for First World problems. *Albania? Her mental health?* The damage was done.

*

For the first two weeks she was on the main ward of the Neuro Intensive Care Unit, a single large low-ceilinged room, whose air felt requisitioned by shock. Two rows of eight beds faced one another like a monastic choir. It was quiet like a monastery too. Occasionally someone would groan – the sound of metal bowing under heat – surfacing for a moment to take a gulp of life back down with them. Otherwise there was the silence of breath held, or screams shorted; no questions, no requests, no jokes, no complaints; perfect patients. There was silence on the clinicians' side too, insulated by concentration, the carefulness one would have with ordnance. With words forsaken they find different ways to express themselves; the patients in the style of their death or recovery, or like her, in the refusal to do either; the clinicians with their preference for tattoos that peek out over the necklines of their cotton scrubs, wisdom texts in occult languages on the underside of their wrists. Like cave paintings the art is not for exhibition, but a deeper need to communicate; that in the absence of any audience, they are still alive.

> *#1 AH, 25, male, resection of grade-four glioblastoma at fronto-parietal junction. Intubated. Single, agnostic, floor-fitter.*
> *#6 DM, 46, RTA at 60 mph. Ruptured aorta, diffuse axonal injury, orbital fracture & left femur / right ulna, possible interruption C5. Intubated. Married, four children, male, Sikh.*

Two of her neighbours on the day she's admitted, printed out for the ward round in a shorthand that tells the nurses in an instant all they need to know. In NICU one could expect three to five road traffic accidents, two to three viral infections, a minimum of four neurosurgical electives, the rest a miscellany – though up close it's never a *miscellany* – of accidental and willed disaster; a botched hanging, an unsuccessful jumper, a faller, a spontaneous dissection in a major artery, an argument between a couple that suddenly took a nasty turn.

Every door in the hospital has a charge, thresholds between health and something else, but the door of NICU is heavier, more guarded, more closely watched. Clinicians dip themselves like believers in a font of alcohol gel and assume the look of crisis. Re-emerging, even the most hardened reflexively lower their head a fraction before those waiting, like emissaries from another plane of existence, still blinking, reconfiguring their faces until the right shape is found. They have come to tell not talk, the message a hit and run – the latest numbers, hedged predictions, deflections – delivered with the authority that knows however much it enhances the error-rate of care, artificial intelligence can never replace them, or recreate that look of experience mixed with understanding that moves between heaven and earth multiple times each day.

★

It sounds fantastical, certainly none of his medical colleagues would agree, but he thinks this milieu will have left its mark in some ghostly form while she fought for her life. Technically she was in a coma for several weeks, but to say that nothing of this place – so alien, so urgent, so traumatic – was registered, is, he thinks, the real fantasy, unscientific even, an adjunct of the same childlike reverie that permits them to think that nobody is watching them, or that she's asleep, or that when she wakes it will be a single discrete moment, feeling nothing for the first time, remembering the weight of the swimming-pool water above her as the last thing she would ever feel; disclosed as she gently comes to, as though in a fairy tale.

Rather than abrupt, deranging, nightmarish; woken, say, by the sound of a ghastly hacking cough, coming from somewhere beyond herself, before removing the gold-tip cigarette from her mouth, flicking it off the building and returning to the serving line. It comes as no surprise to find a tennis court abutting the ambulance helipad on the top of the brand-new Neurology Wing – landscaped with sea sedge, carnation grass and black almond trees – part of the cultural flux, she assumes, that came when they twinned her home town with Vitebsk. On the other side of the court is a seating area where Marc lounges under a parasol with a thatched canopy, umpiring from behind a large margarita, beyond him an infinity pool edging the top of the building: the unpublicised reality of NHS spending, she thinks. At the other end the spry surgeon in bloodstained scrubs and apron, is waiting to receive. The court is a pale, off-white colour, making it harder to pick out the ball. She's seen blue courts before, even an orange one cut into the lawn of a gated home outside Santiago, but never one like this. The surface is not as true as it might be. There are areas where it bows and swells, flatnesses, undulations, generally distending towards the net, rippling towards the tramlines. There are patches of dark sprigs of grass poking through. No excuses, it's the same for both of them. Stooping over her leading leg, the ground doesn't take the ball's bounce so well, barely making it back above her ankle. Her feet give as she twists and launches to meet the toss, forcing her to adjust the overhead and meet the ball in front of her face, lobbing it limply over the net. The surgeon has time to voice a full apology in a thick Montenegran accent, before tucking into his gift, drilling it into the deuce court for an easy point. He holds up his racket, asking for a moment. Taking large scissors from within his gown he cuts out a neat rectangle from the back-court, she can just about make out dark oily liquid under the court's surface. On his knees he pulls the topside of the fresh rectangle towards him, closing off the hole; as he pulls, the court's grosser ripples un-crease, leaving a truer surface. Producing a foot-long needle from within his magic apron, the surgeon threads it with thick nylon cord and sews the sides of the hole together.

It's only then, sitting on the service line, gathering in the view, only then does she understand the court is *her*, that there's no difference between the floor and her feet, both are covered by the same skin; that the off-white playing surface is really herself stretched out, unfurled like a giant parachute that begins and ends at the top of her neck, the sprigs of black grass really fly-like bristles that are part of growing old. Breathing in, all she sees flexes in front of her. It's strangely compelling, this feeling that nothing in body or mind or world is separate, that thoughts are the same as hair and tennis rackets, that there are no means to escape and escape itself can't mean anything. But, she wonders, if this is her skin, then why can't she feel the surgeon's cuts?

'Love-thirty,' says Marc. 'Play.'

Or woken, say, by music; Alban Berg's Violin Concerto perhaps, whose opening arpeggios sound like the orchestra is still tuning up when actually the performance has begun, so too one note will emerge from the noisy throng until she realises it's an alarm calling her from the dream of tennis to a morning in bed. A trapped bee bangs angrily against the window behind her. Something is breathing heavily in the corner. It is night. The bee is silent. The breathing is laboured, a large animal choking. The room is dark other than a terrible orange light coming from under the door. The alarm is a fire alarm. Smoke rises up through the blankets. The fire is her. She waits for the feeling of heat, the acrid taste to reach her mouth, the smell of burning flesh to arrive at her nose.

'Gamma knife please.'

Waking once again to find the room shockingly white and two masked faces looming over her, irradiated by the most powerful surgical light. Neither is familiar and it's also somehow clear that the neurosurgeon is the son she never had, the surgical nurse the mother who died soon after she was born; these are the incontrovertible truths of what she sees. Beeping machines have been replaced by an ensemble of journeyman musicians in balding tuxedos playing Berg, the music the surgeon – a self-proclaimed Renaissance man – prefers to work to. Clichéd buffoon. She can feel the different parts of her own cerebrum as he touches them to the rhythm of the music. She matches the sensations with the changing expressions on the surgeon's face as he saws and cuts and scoops; the feeling of him carving his initials inside her skull, a heart's outline, an arrow through it; forever his, forever hers. Meanwhile the nurse, her mother, breathes for her in the corner of the hospital room, as she had once before, only now sixty years on her breath is loud, bronchitic, the gaspings of a stalker desperate to catch up.

Or she is woken by a gentle 'Good morning.'

(He imagines how each time she wakes things become a little more real as the living nightmare that waits for her slowly pulls focus.)

'Good morning Miss Bella.'

This time it's Robert Mugabe in a nurse's uniform.

'Good morning.' The tone of someone who isn't used to being heard; Camilla Parker Bowles with a Belarus accent, from the other side of the bed,

'*Turn those fucking lights off*,' Bella barks, but only the gurgle of sputum deep underground is heard.

'I'm Dora,' says Mugabe, 'and this is Mo.' Her speech is pull-string, ominous-sounding.

'We are going to evacuate you.' She is a building on fire, people need to save themselves.

'Then we gonna try wean you.' Wean? She's only just been born. The milk from the drip bag is thick, yellow-looking, rotten. Parker Bowles, whose arms are thick and corky as one might expect in a plumber, pulls out a pipe smuggled into her somewhere just below her eyeline.

Bella has the feeling her body is not where it should be since the surgery: one leg trapped at an odd angle behind her head, the other springing from an armpit which is free because both arms have their stems in the opposing shoulder, while her pelvis has rotated ninety degrees so she faces the wall while she lies on her back. She'd had a terrible experience the one and only time she took LSD, a night she thought would never end. That was child's play next to this. Inside her too she senses the surgeon has created new designs: ovaries paired with kidneys, the stomach has been rigged like a bellows to one soaked lung, the other a plastic bag left to flap in mechanical wind. For his *coup de grâce* she sees that he has sliced off her face (saving her nose which appears to have grown, the tip nearly touching the ceiling), though her intuition can't be confirmed by touch, so that it too must be *inferred*, somehow, like the rest of her.

Parker Bowles's hands are not steady and nimble like Mugabe's, an apprentice's hands, fiddling with a smaller pipe until there is traction and the sound of something viscous being sucked out. The stalker breathes more easily.

'That's better. I'm going to raise you up.' Smooth hydraulic noise as the room slowly tilts downwards, ceiling turning to wall, to window, until an outline under a blue sheet comes into view, her that is, only half-deflated. Then the room yaws unevenly towards one of its four walls.

'We gonna turn you, don't want the skin on your back getting compromised,' says Mugabe. Movement without weight, one wall becoming another, white as a cinema screen.

The noise stops, the hush before the film starts.

This is how he imagines it, waking up from a first-person perspective, fragments mixing details of her previous life with its new element: moments of being woken within waking, made more psychedelic by infection, by wildly fluctuating drug regimens, by derangements of posture, by the imagined touch of different nurses, by the absence of that touch, moments which when run together might have the feel of a film – a film that is her life now, a biopic poorly spliced together, with terrifying gaps in between, in which she is moving upwards through the dark unseen waters of her unconscious towards the surface – reality – beckoned by the sound of a ventilator, called by a million alarms towards this second birth, aged sixty-two this time, a Big Bang in reverse where past and future have re-compacted in a present with impossible density, a density she will come to understand – slowly, hallucination by hallucination – is as different from what she had known as it is possible for it to be. And in a corner of the film, like a glitch or a signature, the recurring image of a face in close-up, slowly becoming more stable, holding small neat white signs with marker writing: the date, the day, the place, a name, her name, again and again and again, against the backdrop of mechanical breath, white walls, a remote-control bed, Robert Mugabe, Camilla Parker Bowles, and the rest of the cast in this cheap metaphysical soap opera. But there is nobody there to read the sign, nobody to dispute it, nobody to notice what changes and what stays the same from one whiteboard to the next, like being stuck in the birth canal, until there is something there and the word 'Bella' is read – her name (not her real name, but the name he gave her) – and the same moment she's able to read it, she wonders why the fuck she's being shown a stupid board with her name on it. 'Oriented' is what they call it from the outside – in the dimensions of space and time, in the taken-for-grantedness of reality, the basic what, how and who of being alive, in being Bella, plain and simple; brought back to life, saved from death; the beginning of the end.

Really though, on the inside, where it really matters, the orientation is only beginning. The facets of the new accommo-dation are endless, the rest of her life's work. The cornerstone of her adjustment happens some time later with a grim detonation in her understanding – one part relief, three parts despair – that it's still *her, Bella,* the same thing either side, except her vantage point has been confined to a ledge beyond which the world takes place without her. The understanding is made possible because one thing at least proved indestructible.

'*Everything?*'

'Everything,' said the doctor.

'Lost?'

'Lost. She's lost everything below the eyes.'

'I see.'

She'd overheard them telling Marc at some point, she couldn't be sure when. Which meant her mind – strangely considered 'above' her eyes – had been spared in order to learn of its losses. Not just spared but pristine, enhanced, sharper because unencumbered; the chief executive and only remaining employee of a now entirely virtual business, no infrastructure, no overheads, pure head, the undisputed king of a nutshell counting infinite space. Still the same in essence. Though even on this ledge – from which she sees that Bella has carried on broken but unbroken – her footing is uncertain. She can't be entirely sure she's the same, as she slept through most of it; something could have been swapped easily, like peas under a magician's cup, making what she felt not the proof but the inference of sameness on either side, the illusion of continuity, the contrail of selfhood rather than the thing itself. What knots! But then this voice in which she was now talking to herself – mordant, judgemental, stuck in its orbit – well that was unmistakably her; she should know the sound of her own voice after sixty years.

Empty as a nutshell so it might be instantly filled with calculation, at the mercy of a wild, stricken actuarial urge that would characterise her thinking from waking up onwards: the need to ramify the details of what 'everything' and 'lost' actually mean.

The attempt to map it was incessant and chaotic, like a rat in the maze of her own experience, surgically stroked so as not to learn efficiently, having to navigate obstacles – sometimes sticky, sometimes hard and cold, sometimes electrified – or, worse, absences, passages with no walls at all, or a trapdoor where she gets to fall through herself again and again, and again and again. It meant a never-ending schedule of pain and its reinforcement, in order to fully appreciate the exact shape of what was no longer hers, the un-summable sum of every missing sensation, every disused faculty, every forbidden action.

Like smell: her smell, what she smells like, her armpits, her breath in the morning, distinct from her breath in the afternoon, lilacs in her hanging basket, or lemon bleach under the sink, coffee with honey, chana masala, airports, pine glade in her car, all with the same nose she picks with arms that don't work, whose tip is the only bit of her face she has ever seen unreflected, which she has the habit of stroking when thinking, or not thinking but daydreaming, sometimes anxiously, or running along the cock of her man, part of the rapt underwater exchanges of the sex she will never have ...

Over and over, different iterations, the same algorithms she used at work now applied to her own sensorium. Set the parameters at the worst and work backwards, assays run again and again as they pop into her mind out of nowhere, until the exactness of the catastrophe has been modelled, the new operating system updated with thousands of new synopses of herself in different states that together will form a picture of her trapped in the middle panel of a triptych with nothingness on either side. What pointless labour! All in the name of *adjustment*.

There was really only one simple rule to learn in principle: everything would taste the same from this moment on; a white room whose only light was grief, all the other wavelengths excluded. And it was so simple that it could never be learned, or it could be learned but not endured. And the very worst thought of them all – worst because it was likely the only one worth anything – what if all this thinking wasn't orienting, wasn't adjustment at all, just another layer of denial, an elaborate one, which might look like it was digesting the intricacies of her situation, when really it was just another way of stopping her feeling its simple brutal truths?

★

Certain grosser facets of her situation are grasped more quickly. Like the fact that the stable don't belong in Intensive Care any more than lobsters do, that the absence of drama can't be tolerated. She never sees doctors. The nurses and therapists are on a rota, triaging her care to the 'downtime' between emergencies. Apart from postural tolerance, and the fading hope that she might ever be weaned from the ventilator, there are no goals for rehabilitation. Without goals then whatever she has, whatever she is, is all there is. She registers their disappointment, that things have remained the same, that she remains there, unmoving, for a couple of months now. Ventilator-dependency means there's nowhere else for her to go in the building. She must wait until a space opens up at the specialist hospital, where 'rehabilitation' will mean something again, a fantasy of resurrection sponsored by state-of-the-art technology, the thought of which, they tell her, should inspire hopefulness, and, presumably, deflect their uselessness. The future is therefore measured according to where she is on the waiting list for a bed. She begins twelfth in line, but it's hard to translate the number into a prediction of time; patients might spend a year there or die of shock within the first hours. The anticipation of moving up (a call is placed to the hospital each morning by the clinical lead) becomes the only thread in the endless row of days – for them, that is. For her, he imagines, it's as fatuous as the pop charts.

Behind the glass the neuropsychologist watches the atmosphere grow attritional, the carers turning colder: her unwillingness to meet their hope in the National, to extend herself beyond one-word answers, blinking yes and no, which puts the whole burden of describing reality, her reality, on them; it will pale still further as they grow incurious about how she is, reluctant to describe the view of the town, stop bothering to tell her exactly what they're about to do or why they need to do it. People die in stages; biologically, but in the minds of others too. And things grow on the dying (those minds gorge on reasons for their murder), mushrooming overnight, obscuring what lies underneath. They start to see sullenness, apathy, occasionally spite in her face – which may just be theirs masquerading – so where he pictures an older Bella Chagall looking out through huge violet limpid eyes, they see a crone, head screwed in place, a miser's eye following them wherever they go. A personality with no person underneath. They allowed common sense to harden their hearts – what more *could* they do for her – and then her refusal to change or move – to still be there day after day – to turn monstrous. (It's not to judge, he knows they save countless lives. At the time he looked on and did nothing; only years later will circumstances force him to give her inner life any attention.) The delusion slowly obscured the most self-evident reality that stretched out before them: that there was no existence imaginable more barren and lonely than hers; that the life of a person with even the slightest degree less suffering – beginning, say, at the next vertebra down on her spine – is one that she would dream – could only dream – of living.

Feeling the different nuances of this failure to be seen, and with nothing left of her inner life to savour, only endless ruminations on loss, she is forced outwards, looking with such silent, dedicated attention, becoming more like the crone they see until the situation is reversed: she is the observation station, her large unblinking eyes become windows behind which she takes note after note, recording every facet of her care. Expertise doesn't take long. Soon she knows the rota better than the Sister does, knows the when and how of what they are supposed to be doing, knows each of them by the sound their shoes make on the tiled floor, by the hairspray they use, the length of breath they take before speaking, what mood they're in by the roughness of their hands when they change her – even when they don't know it themselves, even when she has nothing to feel the roughness with, as though her insensate body has become a blindman's white stick, dead but still sentient. Whole careers spent unobserved – unconscious patients behind card-swiped doors – but in the last twelve months the successive scrutiny of a Care Quality Commission audit, a season of twenty-four hours of A&E, the debt that brought the management consultants from PwC in and now, finally, her and her cold observing eye: she is giving them the feeling of what it is like to be her. Reality created over and again like geological strata, one fantasy at a time so that paranoia about getting something wrong in front of her makes them more liable to make a mistake, motivated to get out of there quicker than carefulness would permit, away from the magpie trapped in its cupboard.

12.40
VC-CMV, VT 450 mL, f-12, Fi-O$_2$ .60, + 8 cmH$_2$O PEEP
No change.

The ventilator notes, the number on the wait list – still twelfth – blinking 'no' and 'no' again, are all that's left to say.

5.30
VC-CMV, VT 450 mL, f-12, Fi-O$_2$ .60, + 8 cmH$_2$O PEEP
Alarm attended.
No change.

No possibility of movement. No feeling ever again. On both sides.

Deadlock. It's going nowhere.

8.30
VC-CMV, VT 450 mL, f-12, Fi-O$_2$ .60, + 8 cmH$_2$O PEEP
Bolus given.
No change.

Something has to change, and so it does. Together they will into life an incident, recorded for posterity on an 'incident report' kept at the front of the medical file for quick reference, and which he will position as the hinge in her story, the moment when this tawdry existential thriller stops being merely psychological and leaps into action.

★

The nurse she thinks of as Simulation L387d – 387 for short – looks just-woken, her uniform slept-in, crumbs of sleep unremoved from the corner of her pink eyes. Bella watches, sight sharper by the day, the kind of sight, he imagines, that goes beyond itself: the capillaries around 387's eyes have broken from the strain of her crying, the tidemarks of dried tears line the edges of her cheeks: Bella sees young heartbreak. It's there too in the distracted way she folds over the sheets, the anger with which she punches reset on the ventilator panel. She's been dropped by her fiancé, she's had an argument with her housemate about the shared bathroom, she's sick of the city, she misses the hills that surround her family home.

387 disappears behind the bed leaving Bella relieved not to have to look at her for a moment.

'We'll try weaning again OK?' Word-strings from her predictive text mechanism that wouldn't pass Turing's test. The use of 'we' is part of an effort to lure Bella into a fantasy of agency, to keep her at a remove from the reality of her own life.

'Here we go, OK?'

She doesn't know what's being asked of her or, given that whatever the nurse is doing is already underway, why it is asked at all. The rhythm of the machine changes. In the space between breaths she hears her heartbeat counterpointing the mechanical breath. After a few more rounds the heartbeat sounds mechanical too.

'You're doing great,' says 387. But she is doing nothing. 'Keep going, we're going to try one more time.'

Her heart is quickening, beating twice for each mechanical breath, then three times, as though the two are waltzing, until they are joined by an alarm in a different time signature, turning the waltz to salsa. The tube that comes from her neck is detached, for a moment she is alone, separate from her mechanical mother, her chest rising, something like pain spreading up from her torso towards her head, the ghost of pressure hallucinated into life by expectation.

'I've got you,' says 387. *She's got me*, like it's a space walk for NASA. Most of the time Bella fights to be reasonable but inevitably some unruly thought – fear, regret, anticipation – explodes this, bringing down an avalanche of all that she cannot do, will never do: no tennis, no piano, no cycling, no scratching, no washing up, no touching, no stroking, no fucking, no more avocado or shivering, to never have used a strap-on – the oddness of the order in which they appear isn't without comedy – and when the thread stops, exhausted, ashamed, it's only a moment's respite before the deluge starts afresh, from another angle on the perimeter of the same thing: no ballet, no tickling, no mushrooms, no bananas, no reach-arounds, no reaching ...

'Keep going.'

Without feeling it she can sense a terrible constriction in her long white neck, round her throat's blowhole, her new naval. There were moments, in the early days, when she would really feel something, in all of its aspects, until it fell apart suddenly, exposed as the mirage it was. The world, narrow as she could imagine, could still get narrower, enough to make her wonder if what she had left amounted to any perspective at all. Now three months in, there's just loss. She could plant some sort of flag in that, a via negativa, rely on it to be there when she returned to it, more or less.

Every one of the machines is screeching at the same time in different keys with no obvious signature, Berg becoming Schoenberg.

'Relax ...' says 387. The sound of shallow, urgent breath on the wrong side of her, the outside; the one constant, loss apart, and the only metric that for all its horror allows the rubber of thought to grip the road of time again. With time comes memory and a different order of horror. The bed sheets pulled back, her nightgown ridden up, she looks over her own legs, recalling for the first time the last time she felt them, the last time they belonged to her, on the side of the pool, the heat and roughness of the patio, the shock of the cold water, forced to keep looking on these impersonations of legs – no arms to adjust her gown, hide her immodesty – greying, thinning, losing their stuffing like an old doll, the feet twisting down, arching spastically into the en pointe she could never manage as a child, toes piling up on top of one another, as though trying to fit through a narrow hole, nails black around the cuticle radiating out to dull yellow. She remembers reading about Marie Taglioni, the first to dance that way, and the fans who ate her discarded shoes with sauce; what a way to make your case for love.

'How did she do?'

Mugabe's voice from the door, alarms silencing one by one, the mechanical ventilator slowing.

'Same. Unstable past twenty seconds.'

Memory making the losses worse by creating scale.

'You did really well, Bella,' says Mugabe as she leaves.

'*Help! ...*' Her blinking is clumsy. '*The bitch is trying to kill me.*' She never speaks so coarsely normally. Equally she fears the reverse: what this new reality might do to her past. Her retirement was never meant to be a realising of longed-for futures, but a taking stock of what had already happened. Her life had been what it was: turbulent, dull, vivid, meaningful in bright bursts, then more stably so, even peaceful these last fifteen years. This changed it all. It frightened her, the thought of memory returning with this fate as its terminus; the thought that nothing that went before could retain its value; nothing, no matter how remote, how simple, how unequivocal, could resist being bent towards this, a new gravity that brought all the other fragments of living into its orbit. Either that or, like a terrible lens with one focal length, this catastrophe would see only its kin in what had gone before: other losses of feeling, times where her sensitivity had failed her, where she had failed to love or be loved, like moving backwards down a long corridor where her memories are rooms on either side, towards its final door. What a survey! What a piece of work for her mind to carry out alone, without anyone to discuss it, without a body to support it; the story of her death retelling the story of her life.

'One more thing, before I leave you in peace.' 387's face looks brighter now, enlivened by attempted strangling, pushing her hands into a killer's latex gloves. 'Just make sure your airway is clear.' A different way to choke her. Peeling back dead lips that were once hers, slipping a cold steel hand between a gap in the jaw. Bella watches the view capsize upwards, understanding that her head has been tilted back, mouth opened, imagining those white rubber fingers crawling down her throat, looking for the animal, the one thing alive down there. Apart from the idea of it, and a few splinters of detail, her memory is still just a dull feeling, unable to arrive. Whereas the future has arrived all at once, in a way that couldn't be parsed. They told her retirement would mean all the time in the world, but nobody told her the body would be pensioned off without her, leaving her compelled to watch it devolving back to dull inorganic hospital-issue material. 387's metallic eyes hovering inches above her, the look of heedless intent. She thinks how action starts with the eyes, pilots which the body follows. There had been plans, beginning with moving to the home in Albania, learning the language, swimming, Marc learning to paint with oil. Now it would be one white room after another, the angle of her bed the only variety. It would be watching her body reduce to two dimensions. It would be watching her partner dissemble his suffering. It would be invigilating on all that was wrong with her care. It would be a handful of visitors that she would eventually be persuaded to admit, who would see her once, twice at most, before they could not deny what she knew beyond doubt. It would be thinking endlessly of what her situation was actually like, thinking metaphorically that is, giving herself some sort of handrail to help understand where she was now. None of them came remotely close: like being trapped in a surgical anaesthesia where she felt everything, worse, where she felt nothing but would imagine everything that wasn't felt with phantom acuity. Because unceasing, agonising, raw awareness, one moment after the next without a body to register it is like nothing else, its own metaphor, a thought experiment that can't escape its thinking.

387's eyes narrow as the felt-for quarry is neared, then freeze, pupils widening, broken capillaries leaping out, lips turning thin and white, muscles tight round the jaw, the rictus of a half-smiled scream from which no sound comes out. She cannot speak. Neither of them can. Bella feels her face mirroring her torturer's, the two of them looking into each other across this uncanny valley, as searching, as helpless as drowning lovers. *What have I done?* Conjoined by a bite. Unit L387d needs this paralytic to open her mouth, say something to relieve her, but Bella has nothing to say. It looks like real suffering, thinks Bella: like her own suffering mirrored back. It was just a reflex, the jaw accidentally stimulated by the marauding fingers to clamp down, and nothing to send the signal to stop.

'Bella ... please?' No more 'we', her voice finally devoid of sing-song jauntiness. 'You're biting me.'

Bella is not biting, but the ghost of her might be, an emissary from her unconscious delivering this nurse something of her own pain.

'Please.' She lives alone, she hates the city, there is nobody to come home to.

'Relax your jaw.' There is no hope, knowing Bella has nothing to relax and nothing to relax with. Bella feels her pain and is helpless to respond. The tears are streaming down their cheeks, heartbroken by the other, love a singularity, beheld in every face you meet. She didn't mean it – look at those eyes, innocent, unclouded, sympathetic even. Not a robot but a girl now, in agony. Finally someone to understand what it's like to be her; a senseless, feeling thing. And with that, a moment of relief, of satisfaction even – though it's hard to admit it to herself – that she can still cause something, still communicate, even now. She didn't mean it, did she? Tears meeting blinking tears, agony meeting joy, joy and guilt, two realities touching one another for a moment, but not quite. She imagines blood pooling in her mouth, real human blood trickling down her throat, her first food, new food for a new her, the feeling of her teeth meeting bone.

Help.

'Help.'

★

The incident gives her new life, as something which bites back. They are not just imagining this. The situation confers on her new powers; the talent for repulsion, the gift of the sick, the crude, the old, confirming what they have long suspected: the crone is after them, the magpie has just been biding her time. A sequence is set in motion. Next she sets

Marc on them; loyal, gentle, courteous Marc from Troon, whose hair has grown long, whose clothes stick to him, who always carries a cup of tea he forgets to drink. There's menace in his voice when, a few days later, he makes a complaint about one of them being late for her shift, then another about one of them being 'bullish' in her bedside manner, and another, the same day, about the one who only ever half-cleans the *fucking* feeding tube. They will not tolerate such accusations. They will not be spoken to in this way. In the days that follow hours go by when there is nobody to change her pad, cut her nails, mollify her cracked skin, nobody at the observation window, nobody recording her at all. She is left alone, the new queen of NICU, like Louise Bourgeois' spider, all of them caught in her psyche, struggling to break free of the million-threaded web of her loss. (Writing about her years later the neuropsychologist feels it too, as though his awkward, angular airless sentences – each one a failure to see more – were somehow caused by her, an extension of her total loss of feeling, as though they are the only way she will let herself be imagined.) Until enough is enough and one afternoon two of the nurses turn her bed round – not easy in the space available – so that those unflinching eyes are made to look out on that rainy northern town which is on the same latitude as its twin, Vitebsk in Belarus. 'Plenty to look at,' is what they tell her, closing the cupboard door, then putting that cupboard in another cupboard pitch-black, blocked off by a rock, in a cave, at the bottom of an empty swimming pool on the deck of a ship on the bed of a fathomless ocean.

★

Occasionally there's a doctor whose face remains the same either side of the door, who won't recognise the threshold, or any other fatuous superstition for that matter, so that on entering there is no discernible change in her step or expression (when she was a child she dreamed of one day looking like the profile of the Queen in bas-relief) as though the passage from life to death itself will be nothing more than a convention that need not be observed by everyone. Such purposeful effrontery, balanced on high heels that click-clack through the silent ballet of green scrubs, has come from an entirely different order of reality altogether, a different genre one might even say: one that intersects with Bella's momentarily, but decisively. Nothing – not house fire, school shooting, train derailment or killer virus – could change her cadence or course, or tamp down her gunpowder-blue trouser suit, or threaten the direct, unhushed, incurious tone with which she asks the startled nurse for directions to Bella's room as though she were asking a groundsman for directions to the visitors' cafe at Windsor. Because Dr Elliott, monarchist, one of two female neurology consultants in a department of forty, one of three who didn't go to Oxbridge, unique in her lack of research publications (considered time wasted when one could be saving people), is in the business of generating her own reality. When Dr Elliott clears her throat and opens her mouth, it marks the beginning of the story, distinguishing – in a voice that (unless slackened with Prosecco) never hints at the Glaswegian housing estate from where she came – what truly matters from what is merely incidental. This feel for the true nature of things, however fraught or unstable the environment, means she doesn't have to say much for the future to become an instruction, tasking the rest of us with making sure it arrives on time. If her head nods when, after introductions, Marc begins telling her about the difficulties they've been having with the nursing staff (taking hold of Bella's dead fingers as a way of grounding himself), then it's before understanding is possible; rather the nodding is a way of keeping his words at bay. Dr Susan Elliott – never Susie – is not concerned about impression management, not in the business of hospital liaison. She always has all the information she needs, she always knows exactly what sort of people she's dealing with. Now she's here, standing in front of them, ready to remove any obstacle that stands in the care pathway, dead or alive. Still nodding until she can bare listening no longer, she interrupts to correct him: her absence as Bella's named consultant was not neglect but good practice; from the medical point of view everything has been satisfactory up until now.

Is this a joke?

'It's important to let the dust settle.'

Dust?! ... Dust is all that's left.

'It's what we mean when we say "watching and waiting".'

Like I could do anything other than watch and wait and piss and shit and cry.

'In terms of trauma it's very early days, the picture is still unfolding.'

The picture has unfolded, finished, monstrous.

He is there to see this, part of the consultant's attending retinue, saying nothing to dilute her effect. (He still feels shame like hot stones on his chest to this day.) He's read the Literature online, papers suggesting that the nature of early interactions between medical professionals and spinal patients is critical, predictive of long-term adaptation, well-being, life expectancy. He wants to say that even though they think the damage had already been done, it's being done all over again.

'Patience is the key in the aftermath of emergency care, especially when the brain is involved.'

'We were told that her brain was fine,' says Marc. There's dry toothpaste in the corner of his mouth.

'Though the main bleed is at the top of her spine, there may have been a separate cerebral injury.'

'Separate ... How?'

'Oxygen deprivation.'

'Is there anything to suggest that on the scan?'

'Scans don't pick up everything, Mr ...' She doesn't know his name.

'What evidence do you have?'

'The experience of decades. I don't mean to be pessimistic, we just need to be careful about making assumptions given your wife can't communicate for herself.'

'She's not my wife. She can hear you.'

He imagines Bella digging her nails into the back of Marc's hand.

'I know she can hear me but that doesn't mean she understands in the same way as before.'

He imagines her flaying the consultant's regal profile.

'We can test it, there are ways: yes/no questions, picture sequencing.'

'You know who she is? What she does? How qualified she is? She ran a department with a hundred people under her, most of them with PhDs.'

'She's the cleverest person in the building I'm sure. But even if her intelligence is intact, and I'm not so sure, there's her mood to be considered; it can have a dramatic impact on her thinking and her behaviour.'

'What *behaviour*? Look at her.'

But Dr Elliott is looking at Marc; the mess of his hair, the terrible tiredness in his eyes, the toothpaste. 'I'd like to have her reviewed by psychiatry. Strokes can trigger terrible depression.'

Depression? Is that what this is?

'You're kidding? How could she not be depressed?'

'Depression is a complex mental illness. And in its own way it can be contagious for loved ones.'

Now I have a complex mental illness sitting on top of a body that doesn't work.

'Isn't that right?' Dr Elliott turns to the neuropsychologist, him, to back her up.

'Well, there's grief, the reaction to loss, then there's the catastrophic thinking as accommodation takes place.' His leg won't stop tapping while he speaks, making him look like he's relishing the opportunity. 'Then there's also organic changes to the brain chemistry.' Really it's guilt, anxious guilt.

'There's her history to consider,' Dr Elliot interrupting, as though he wasn't there at all.

'What do you mean her history?' asks Marc.

'Her predisposition, the nurses have reported significant resistance to care.'

No confrontation or jibe or denouncement is sufficient to draw the Irish to the surface.

'The nurses have? Resistance to what exactly ... ? I want to talk to you about the fucking nurses.'

Just the look is enough to push his words back on him.

'There's a general feeling on the ward; always a sign something's not right when the team stop volunteering. We'll get her seen by Psychiatry.'

The royal 'we', teamwork, the sum of knowledge; erecting a fortress around the truth, with the patient outside.

'She may find it helpful. You both might.'

I will eat you alive, all of you.

'Very good to meet you both finally.'

That was it. The lines were drawn and they couldn't be undrawn. They never saw her again.

★

Down to sixth on the waiting list. According to Psychiatry she's 'awkward', 'controlling', 'impossible', 'paranoid', read into the eyes which grow duller by the day of an ageing lady who doesn't move or speak, who has been as distorted by their attention as she had previously been overlooked, a torturous living koan: something that can't be cared for but demands constant tending. A multidisciplinary meeting is called between the factions, which isn't easy, asking for ninety minutes of intensivists' time when they are used to thinking and feeling that only moments count. Dr Elliott can't make it, nor can Bella of course. Chairing it, one of the consultant anaesthetists opens by speaking of how much he admires Bella's resilience, her stoicism, her courage – she had done nothing, she could do nothing – as though he's forgotten where he is and is delivering a eulogy. Each therapeutic speciality takes turns to feedback, intoxicated out of nowhere by their own confected outpourings. The nursing rep, who's not smiled in all the time he's known her, asks Marc questions about Bella's taste in music, her favourite colour, the places close to her heart – as though after four months they might begin again, starting with who she is. They're not used to getting to *know* their patients, of having someone so 'well' to look after ('Not "well", but you know what I mean'). For a moment Marc can't resist this fantasy of starting afresh; who could, in the circumstances? He concedes it's possible his dear darling has become a little more stubborn, less tolerant – uncharacteristically so – a disloyalty he immediately checks by adding that it makes sense, as a response to her powerlessness. Gestures from both sides, however self-deceiving, still a therapy of sorts. The session ends with a commitment to build on this with smaller weekly check-ins. But the meeting a week later was different, only two people turn up to meet Marc and neither had been at the first. They read out notes and apologies from absent colleagues, stuttering their way through updates written in a hurry. They can't have read them over before, otherwise the suggestion from one of their absent colleagues – that the observation window might be made into a one-way mirror, for the sake of their privacy – wouldn't have been read out. The meeting's abandoned halfway through; a dozen emergency admissions from a coach crash on the ringroad. A few days later it's reported at the morning handover that Marc asked one of the junior doctors about the procedure for requesting the ventilator be turned off. 'Over my dead body,' is all Dr Elliott says.

There are guidelines about withholding or stopping treatment in the event of catastrophic injury – this was a while ago, things have changed since – but they are only guidelines.

Without the support of the lead medic it's impossible to do anything legally. Bella can challenge any prohibition, but to do that she would have to prove that she has the capacity to make decisions in her best interests; that was why Elliott had insisted on a mood assessment at their one meeting. Ten minutes with a liaison psychiatrist on a daily locum contract that depended on Elliott's signature for its renewal, a refusal to blink to questions, and Marc's 'uncooperativeness' were more than enough to deny her the capacity, warrant a new diagnosis of psychotic, suicidal depression, and have an SSRI added into her drip.

She wants to die; it strikes right at the meaning of NICU. The ward turns on its axis; instead of more or less conscious evasion, finding any reason to avoid the Gorgon's sight, the team organises itself into a surveillance op. Marc's visits must be accompanied, ventilator alarms are responded to instantly for the first time in weeks, pointless therapies and nursing care are readministered promptly and solemnly. When she isn't receiving treatment a rota of staff are assigned to the obs station. In these ways they turn her back into an emergency because it is the only thing they know, reclaiming power, renewing purpose in the face of something that constitutes a greater existential threat to them than the stable, unimprovable 'living death' of these last months: this is *life* support, not Dignitas. They could kill her for even contemplating it. If she only knew the fight that countless others had been through in these very same beds to stay on the right side of life, if she could see the flowers, and cards in the staff room, the overwhelming thankfulness; National Treasures is what they are. Of course the effect of their surveillance is to make an impulsive assist, a *crime de passion*, all the more likely.

Even now, several years after the fact, the neuropsychologist is staggered by the thought of murderous nurses around a half-dead suicidal patient, how things could unravel in that way. At the time he was caught compulsively in its orbit: cancelling numerous clinics with other patients, skipping departmental meetings, swimming in the middle of the day to clear his head, watching her, watching them, but mindlessly; the drama's grammar shepherding him forwards towards its climax. He sees that she's becoming less readable, more sphinx-like. He imagines how eventually one must arrive at a place where all the ruminations dry up or turn into the same thought, as white and empty as the room she's in. Is that where she is? Might there be peace there, the experience of consciousness itself, pure, without concepts, prior to the body, even her head dissolved, thoughts trailing off before they take any shape, seen for the riderless horses they always were, a place where just seeing is enough to disappear them instantly; finding herself nowhere and everywhere, a limitless plane where the very best and worst, the most intense joy, the most terrifying pain, are instantly equalised.

She is not there. Instead, he imagines something more visceral, being grabbed by something from below, pulled down into the depths of her earliest, most signal memories: the first person that overwhelmed her with the need to be a certain way, to live according to a will that was not her own – her father's. And like an artery, once nicked, memory arrives in torrents, carrying her away from unbearable thoughts of being murderously observed by those bent on keeping her alive, delivering her instead into a gentler psychosis, beginning with that childhood holiday, lying in that tent, thinking the bears were waiting outside, watching her. There are no bears in Wales. Rain had kept everyone inside for most of the week, squalling in the morning out of a soot sky, more fitful spatterings through the afternoon, so that by then a kind of helplessness had been perfected, the day's spark snuffed. Nights when it drilled against the canvas were different, like a rejuvenation, the noise so powerful, so close, sleep was impossible. Instead, with eyes closed and her father safely asleep, meaning her imagination is her own again, she can extend herself outwards, feel herself just as acutely in the space around her as in her own body, until she reaches the limit, her skin blending with the tent canvas so that her outside is outside, and she can feel the rain landing on every inch of her, feeling everything again as though nothing, not a single drop, could fall unfelt.

Six years old, told without ceremony she was too old to share a bed. Even at that age she had done it more for his sake. Theirs was one of the larger tents, the extra space welcome given there were five more or less strangers under the same canvas roof. It had a kitchen with a two-ring cooker and a cool box, next to that a large dead area which was variously commissioned as a dining room, a card school and a boxing ring, then two separate sleeping areas either side of a canvas partition. Her father would be next door with his latest girlfriend Zoe, marking an abrupt relaxing of the pretence of friendship, leaving her to share a bunk with Zoe's kids – Max and Eddie (or was it Davey? They weren't the only siblings she'd been asked to adopt at short notice in the few years since her mother's death) – top to toe in the bottom bed. Even without the rain the week was less a holiday than an experiment in social psychology, like the ones they were conducting in American universities round the same time, where groups of people are put under exquisitely designed duress to prove that individual attributes – decency, loyalty, courage, kindness – are kites in the wind of circumstance. This was an experiment without consent, without ethical approval, without apparent end; three days of solid rain, with more expected, had made monsters of them all.

The participants had met briefly, uncomfortably, once before, at what was then the only Italian trattoria in West Yorkshire. Across a table with fresh red and white cloth, its own foot-high pepper grinder, she listened to the boys boast about killing blackbirds with a catapult, her horror turning to incredulity at the ridiculous accents they summoned to order '*vaniglia affogato*'. Were these the first inklings of an astringency she would feel around all men, that would last to this day, with the exception of Marc? Pembroke was the second summer holiday, but the first that she could remember for herself, confident that the experience was hers. The one before in Cyprus when she was still a toddler had disappeared altogether, apart from feeling the wall of heat and hearing the horrible, relentless sound of angle grinders cutting out new hotels, and even those details, however visceral they might feel, were more likely hallucinated from photographs or implanted by her father's repetition of half-drunk anecdotes to the point she couldn't distinguish them from her own. Wasn't her whole life like that to this day: half belonging to him? That was why the memory from that second holiday – the unshared bed, the feeling of space while he slept, the fantasy of feeling every drop of rain on skin that would never feel again – had found her now; the very end of independence meeting its beginnings. The sheer brio of her younger self, springing up from her seat in the bistro to announce to the group how killing a blackbird was the darkest thing, the work of a black heart. That wasn't all: their liking for coffee poured over ice cream was 'fake' – coffee itself was foul-tasting 'fakery' by which adults marked themselves as old. These *opinions*, these *preferences*, *this vocabulary*, where on earth did they come from? She didn't even know she had them before she stood up and tried out this new tone, these new words. The whole table fell silent for a moment, not sure what had just happened, measuring how much offence to take, while she wondered if other words existed – a saying, a ritual, a spell – to unsay what she had just said. Whatever it was, wherever it came from, this defiance of hers wasn't strong enough yet to be stable, would give way too quickly to a deeper, desperate need to belong, even to a family of strangers, searching out one of the boys' reluctant hands to hold on a soaked walk back from the local pub, letting slip 'my brothers' as she told another young girl she met briefly in the campsite showers who was who.

In her hospital room a stranger is talking to her from a lightweight wheelchair. For some, life is so neurotically a story or memoir that it needs to be retold constantly to have any integrity at all. The one thing Bella knows even before she settles on the meaning of what the young, uninvited woman is saying, her gesturing hands protected with the padded gloves of a para-athlete, she knows she must resist her. This is what real *life* looks like, is what the stranger's attitude says, this here is the difference between you and me: *the running shorts, the carbon-fibre wheels, hard triceps, the passion – a body returned to its rightful owner.* The story launched at Bella, as though it's all she's ever been waiting to hear. The details are unimaginably cruel – a stray bullet in a neighbourhood gunfight entering her nine-year-old spine – but to Bella this visitor who so insists on being real, is more hologram than flesh and blood, less real than the memory of this strange temporary family she had found herself in more than fifty years before. Her horrible flannel pyjamas with printed toy soldiers, crying face down in her bunk bed, the musty smell of the pillow, the rain lacing their temporary home, then her stopping suddenly and looking up in a moment of self-consciousness, to see the boys crying every bit as fulsomely as she, and only then does she hear the much more upsetting adult wail – Zoe's – from next door. Most memorably of all, her father moving between these tearful *tableaux vivants* – trying to mollify each in turn with tailored bribes, distractions, promises of a tent detente that wouldn't survive the evening. Politicking like this didn't come naturally to him; calmness, concern, could only be impersonated. And here, as the social psychologist would predict, the situation militated against him, by the simple equation that attention to one meant neglecting the rest. In the end, which was quick in coming, he exhausted his reserves spectacularly: 'This is the fucking Ebbw Vale of Tears', exiting, only to be thwarted by the jamming zip-door.

The stranger's hands chop at the air. Chantelle: her name like her syntax like most of her culture is American. Really she's from South Yorkshire. *Marathons, law degrees, her own NGO, ambulatory boyfriends, filthy weekends in country hotels with wheelchair access* ... Whatever she says it can no longer get in the way of Bella remembering how clumsy her father's hands were as he fought the zipper, remembering also the thought that from the outside the struggle might look like a Wendy house with a grizzly rampaging through it. Here the bears have opposable thumbs, this one pulling a neat silver-blade from his pouch and, to his teary audience's amazement, cutting a bear-sized hole in the tent. The boys joined their mother at the new entrance to watch him walk through the rain to his car, get into the driver's side, slam the door shut. The one called Max was laughing which gave permission to the others; there was something just irresistible about the scale of the sulk, the way he loped and pawed and snorted, enough to forget their griefs and settle into being his audience. Only she remained crying, wrapped in a towel, not yet seven, too young to get the joke. If only she could summon tears as easily now, like a skunk's spray, to drive away this lady on her carbon throne averring that *nothing gives her a bigger buzz than seeing the light come on in another spinal's eyes.*

And as he watches them through the glass, picking up Bella's reaction from the tiny cues and tells in the top half of her face, and the little spikes in the rate of the ventilator, he remembers reading a study in the amputation literature, supporting positive therapeutic effects of bringing together patients with similar presentations, promoting identification and resilience through shared narratives of suffering and over-coming. But all Bella sees are differences; all Bella hears in this young lady's anthem of hope are the strains of a particular species of magical voluntarism. And he remembers how the study found a significant statistical subgroup where the affiliations were counter-indicated, the amputees forming their own suicide circles in the toilets of the acute ward.

'Look, it's not a choice to live or die, but lots of choices, and they're made one at a time, starting right now.' Chantelle lets her words hang in the air for effect.

'Get out.' Bella has moved her hand and used it to take the tube out of her neck, plumbing the hole with her finger: 'Get your life out of mine immediately ... OUT!' Her voice is powerful, much deeper, more masculine, than he'd imagined.

He was the only one who saw the miracle of her hand moving, the only one able to hear her order, the one who couldn't bear another moment of Chantelle. She put her business card down on the table next to Bella's bed, spun the sleek chair on a dime, and scooted out of NICU across a floor as slick as a basketball court.

Bella is left to herself again, in the silence between mechanical breaths, reminding her of that other silence nearly six decades ago; the one that went on too long.

This was not funny. This was not a show; her father meant it. The rain on the windscreen. The murky figure not moving inside the cream Rover P5, her favourite car from childhood. Still they were laughing, with too much force; like so many people's laughs, an excuse to release every other unexercised feeling, dating back to God knows when. Really nothing could have been less funny, because there is something about a lone man in a motionless car that signifies death. When the car's engine did start, all three litres of it, the window wipers furiously beating back the rain, the laughter died out. Her relief at the sign of life immediately gave way to another thought, worse than death: he is going to leave her here with this new family.

★

Down to three on the waiting list. He drives with Marc down the M1, a hundred miles to junction 22, then across country for an hour. The radio plays a song about a shepherd who dressed as a sheep to look after his flock. Neither of them has passed the city limits for months. Once they leave the motorway there are endless roundabouts, lay-bys with men in huddles by their lorries drinking tea, smoking. Pylons skewer fat clouds. Southbound roadworks lock the traffic in one lane for several miles. Motionless, tiny drivers' heads, still, facing forwards. Then briefly through the trees white chalky hills. Marc talks a little about her, their life together, what he knew of her before they met, the afternoon of her accident, bits and pieces in whatever order they fall out of him. (This is the material the neuropsychologist uses as his starting point years later, when it comes to reimagining her.) He won't prompt any more than is offered, Marc is a private man. Still, not as private as she would want him to be. He offers to change the music. Marc says he doesn't mind, then 'None is also good.' The sun breaks out for a moment. How tired Marc must be, he thinks, of encouragement; the effort required to envision futures for them that stay on the right side of delusion, to tolerate her need to shoot each one of them down with a blink of her eye, or just a look; how, like her, he must have had to keep on accommodating, updating his hopelessness as the reality of her situation grew more pronounced. There are speed restrictions, red triangular road signs with children playing, a bad drawing of a stooped elderly couple, deer outlines. They see none of those things, just the signs for them. He wonders what if anything is still riding on this visit; at this point, as close to the future as it was possible to get, there was almost nothing left to invest with in the promise of technology, in the notion of a 'national specialist centre', in 'expert rehabilitation'; after all that had happened the only force it had left had depended on being remote, a fantasy, a distraction, which diminished as she climbed the waiting list, and had turned to virtually nothing now that they were less than fifty miles away from seeing the idea of it become the thing itself.

From outside it looks like another service station; one storey, dun brown-brick, built in the 1970s to commemorate the future. A woman, a volunteer called Jane, is there to meet them and show them round. She begins on the history of the place before they've got out of the car park: '... opened by the Queen Mother ... some of those C-spine patients are still going strong five years later ... one big family really'. Marc can't stop looking at the buildings, under the low, grey sky, as though all his questions are answered by it. The neuropsychologist excuses himself from the tour and takes a seat on a grass verge overlooking the hospital, drinking tea. Below him children in wheelchairs chase a ball across a football pitch.

'They call it dwelling,' Marc explains later. 'The screen has a light trained on the eye muscles so you can choose a letter just by staring at it.' He manages to lose himself impersonating the jaunty explanation he's heard a few hours before. Meanwhile she's dwelling on him – dwelling is one thing she can still do – waiting for him to tell her the truth, but instead he tells her how, if things go well with the tracheotomy there are voice synthesisers that can work with minimum aspiration, like Hawking's, 'like whoever you want to sound like, like you ... Depending on how much we get back ...' *Get back*? He has swallowed their pill, uploaded their language instantly. But being reassuring and hopeful doesn't suit him, turns him a little wild; hair rising in gravity-defying wisps above his head as he talks about smart houses, shirt half-unbuttoned, yoghurt on his collar. This hope is for him now, for him alone, she still has that job, even if she can't do a single other thing in this life.

'I was thinking how the pool might be customised into a hydrotherapy unit, we'd just need a lift installing.'

Right now dwelling on what he's telling her is too much, she has her memories to get back to. The sun did break out one day. It must have been a weekend or even a bank holiday because the small remote cove, an *'objet trouvé'* – her father was shameless in his readiness to wield any of the three dozen French words he knew – they spent hours finding, was less deserted than he anticipated. In fact it was crowded, cartoonishly crowded, not so much 'found' by everyone else as couldn't be lost. As there wasn't a strip of sand big enough to lay out a beach towel, he picked the spot he wanted, and asked their new neighbours to *shove* up, helping them on their way before they had time to fully consent. Still so alive inside her; the hundreds of embarrassments like this one, stored somewhere in her body, felt on his behalf, felt because he couldn't. There was such urgency around him, in everything he did, however mundane or nondescript, even raising people from their sunny drunken slumbers to move a few feet that way. She understood it now from this terminal vantage, the urgency was a reaction to his own insentience, that he couldn't feel his force sufficiently. His body – big, bearish, brutal – on which he depended to crash through life, to do his bidding without question, was also too remote to hear back from, leaving him with the reactions of others as the only way to estimate himself at all. And where were these others? Zoe would walk or run or swim for two hours at a time just to abscond, because to sit next to him, even in silence, was to be irradiated by his restlessness. Even then, that was her six-year-old's intuition, watching the woman she never got to know, who would have been decades younger then than she is now, with her head down in the water, smashing her front crawl through gnarly surf, as though there were times when any buffeting, any trial, was more life-giving than her father's company. Is that what had done for his wife, her mother? What might get mistaken for exuberant or headstrong was really much more threatening the closer you got. Abandoned himself, he abandoned her to finishing her delicate mandalas made of shells and seaweed while he prorogued the boys into playing beach cricket. Bella knew Zoe's mind because she already felt the sheer sensory relief of having space again, intuited that for all the textbooks that might be written attachment is no more than geography, how far you can bear someone being, but also how close. Having him over there on the wet sand with the boys, milking consecutive boundaries off weak, tired deliveries, never giving the solitary outfielder anything like a chance to keep him interested, was heaven for them both. (The neuropsychologist can feel the freedom her mind finally affords her, the memories coming so easily and unconstrained; feel it in the unbridled way she is finally allowing him to write about her.) Her father would joke with her in the years to come that a tear escaped him when the nurse told them they would be having a girl. It wasn't a joke. Pyjamas with toy soldiers. It was another of his preferences that she could feel more clearly at a distance of years, now that the rest of her feelings had been sheared off. And here he was, in revenge mode, shamelessly cheering himself for a huge square cut into the surf that meant a miserable swim for poor Max who hated water. She never saw Max again after that holiday, but decades later she wrote to him as Conservative Member of Parliament for Elmet, asking in as friendly a way as she could for him to lobby against wind turbines outside Wetherby and receiving a polite reply from his secretary that the matter was under exploration. By then her father had died of bone cancer, secondaries from the prostate. Young, they said, but still a few years older than she is now, his body much less robust than it had always appeared. Not a joke; a tear because a child, any child, was on its way; not a joke because he didn't need dependents, what he needed was accessories, stagehands, fielders in his own drama. Is that what killed his young wife, her mother?

Really what she remembers most keenly from that afternoon was the *effect*. The details would have been filled in later by one of his stories. He would tell it as a cautionary tale, used whenever she threatened to flash her disobedience at him, of which this had somehow become the mythic origin. But he couldn't change the effect, that was hers alone. He had launched a huge six over extra cover, his third in succession having twice promised not to. Max had gone after it again, wading into the flinty cloud-covered sea up to his waist, unable to locate the tennis ball in the swell. He could make out his mother, though, or he assumed the black dot half a mile further out, about to round the craggy headland, was Zoe. Resigned to wading back empty-handed, he watched the whole beach rise to its feet and retreat like a slow single cove-wide wave, a giant's shadow on the move, except for one still point, using his bat to point furiously in the other direction. From the beach, what had seemed from the sea like terror at an amphibian landing, or a second Vesuvius, was a bothersome, drunken scramble not to get wet feet: comic groans, children pleading to return to rock pools, wives lambasting husbands shipwrecked by booze for not helping with the load, except that is for Bella's father who had forgotten the ball, and was running the length of the beach screaming as though molten lava was at his back – that's what love can do – frantically searching under blankets, behind hampers, to pick up a child that might have looked like her from behind. The water forced him back, funnelled in with increasing force as the cove narrowed. Apart from one or two of the more drunken ones who in that sometimes British tradition of Canutian defiance had elected not to move at all, sunbathing in two feet of water, the rest of them were by now traipsing towards the car park, or fighting each other for space on the remaining postage stamp of dry sand. Eventually he found her under a deckchair eating an ice cream bought for her by a kindly-looking elderly lady knitting the beginnings of a pullover. 'Never mind that.' He threw the ice cream into the sand and snatched Bella up into his arms. The elderly lady, all kindness gone, stood up, holding out a knitting needle. 'Do your job, then.' They were staring each other down so poisonously, Bella thought he might hit her, or the old lady might puncture her father's chest with her needle. She hadn't intended this, it was horrible. But it was also a relief, that much was written on her ice-creamed face. There were shouts from above; the boys were waving from the cliffs. The moment had passed but not for her; he had left her and then he had found her, just. It wasn't itself a proof of love, but it made a case for it, didn't it? The details might be totally wrong, but that didn't matter, his story didn't matter. What mattered was the effect and she remembers that with psychedelic clarity: the sense of her self popped out of nowhere, a terrible miracle in its own way, separate, alone, and yet able to knowingly be the cause of something outside of herself – a small thing, which contained life and death – all that for the very first time. Though she couldn't characterise it in that way until decades later, in her thirties, when she came across a poem about a young girl in New York in 1918 – more or less the same age she was on that holiday – and the sudden origins of her self-consciousness while flicking through an early *National Geographic* in a waiting room at a dentist's office. For Bella there was no world war, no sound of a dentist's drill to traumatise her, but something much more familiar; abandonment then discovery, being a *thing* that those things could happen to, that could cause those things to happen. If now was the end, the ceasing to really cause anything ever again, the very last abandonment of them all, then this moment was more or less the start; both of them made brighter by their proximity to her not being there at all. How funny, how wildly strange that in her childish, psychiatric reverie she should chance upon the clarity to resist the seductive, reality-less inkling – the same one that her nurses had come to believe – that she had brought this whole calamity on herself. More than that, she's left with a deep and reassuring sense of those things that can't be thought; the sense that even when she could move and feel and speak and laugh, the means by which she did them came from nowhere and remained utterly mysterious.

★

She is number one for weeks, then a phone call late in the afternoon says a bed at the specialist hospital is available. She needs to accept it by the following morning or she'll lose it. She is dressed by nurses while Marc collects the few scraps of memorabilia used to make her feel at home: the Chagall painting he'd brought in, a photo of her as a young girl with her father. The room returns to itself effortlessly, in a matter of minutes, like she never existed. The lights have been dimmed on the main ward as she and her portable ventilator are wheeled past the sixteen beds of men and women who could so easily be mistaken for sleeping. (He remembers holding the thick doors open, seeing for one last time, those eyes, that hair, all the things that were real coming and going from the portals of her face. It wasn't that she reminded him of someone then, but that someone else would come to remind him of her.) He imagines it will be dark outside by the time the ambulance gets on the M1, so she won't get to see the changing scenery as it takes her away, five months after it had brought her, siren still flashing.

Marc will go back to the ward several months later, after Bella has died, to donate two sun loungers – bought for the holiday home – 'for those rare sunny days'. There is a rooftop terrace where long-term patients, nurses on a break, or distraught family might take a breather. He looks altogether different, clean-shaven, smart dark trousers, an ironed green shirt, relief, but also determination. He wants to make peace with her time there, to show that hard feelings had softened and hide the other harder feelings that replaced them. Mugabe, Simulation L387d, Parker Bowles and the rest, they listen with shocked faces as he tells them without a hint of drama about how she died of asphyxiation after a few weeks, the nursing staff not responding to a ventilator alarm in time; it looks like shock anyway. There is an inquest pending.

Before he leaves Marc takes one last look; the bed's been pushed against a wall, old files scattered over it, yellowing computers and their monitors taking up the floor, a broken-looking ventilator lying on its side.

17.30

VC-CMV, VT 4500 mL, f-12, FiO_2 .70, + 8 cmH_2O PEEP

No change.

Alarm attended.

Marc closes the door behind him and draws the curtains across the observation window.

One last look, a final survey, his queen laid out in state: her beautiful toes, still knock-kneed like a young girl, tummy the colour of milk, shoulders that pull her through the water ... feeling every inch for her, because there was only one nervous system between them.

He gets into bed next to her, laying his head on the sea of her chest a moment, listening; the shushing like waves over shale, feeling the body's tender warmth, wasted unshared all these months.

'Shall we?'

'Yes,' she blinks back, and 'yes' again.

He removes her gown, pulls out the tube from her neck, and all the other bits of plastic and metal that have been plugged into her body. Ignoring the terrible noise of the alarms he wraps her carefully in fresh sheets, like a shroud, lifts her up in his arms — she's already light as a feather — and carries her to the window, holding her so that together they may look out over the town hundreds of feet below. Standing now on the thinnest of ledges, he can feel her heart beating faster at the lack of air, at the thrill of it all, and holding her as tightly as he can he steps out ...

They float upwards, held by love's vapour, the air underneath. Below, the higgledy-piggledy town, pylons, scrapyards with high barbed-wire fences, mosques, grey except here and there a house or a tree astonishes them with its colour, but not the next or the one after that which are hopelessly dull by comparision, though don't be too quick to judge as they too have a red window, or a yellow door or chimney stack; as if to say – the magic is even here, if you look with care. Flying again like they used to. It's impossible to have been wild once and not want it back. A stray dog is barking joyously at the sky. The bed sheets wrapped around her look like a dark blue dress flapping in the wind, her white neck brace a lace collar. She rests her head in the nook of his brilliant green shirt, holding her hand out to feel the air as they sail over the town.

Over the Town, Marc Chagall

Davis

On the second day

A lone window, too high to see out of, a single metal-framed bed bracketed against a wall, in one corner a sink next to an unscreened toilet, in another a heavy armchair whose fire-resistant finish sticks to skin: standard psychiatrica. Except Davis prefers the cold linoleum floor, legs folded under effortlessly, naked except for the bed sheet tied around his waist, head turned, like a saint, towards the high window waiting to receive its light. Nobody before him has ever sat this still, the silence pouring out of him with such intensity that he's still there now twenty-five years later; a scene without exit.

Almost nothing was known about him, just innuendos, swarms of them moving like currents through the building. There are patients who need imagination to bring them to life, then there are those where the fantasies come so thick and fast the work is to stop them. The silence, the stillness, the anonymity; nothing could have been more exotic in a place like this, nothing more suggestive at that moment in the neuropsychologist's life. The feeling that whatever Davis stood for belonged to ancient planes, where the land has flattened over hundreds of millions of years. Also the feeling of its opposite, that being in his presence that morning was the rarest of events, when something genuinely new has come into the world.

*

The half-empty Air India flight had arrived in London at 6 a.m. the day before. It wasn't until all the passengers had de-planed that flight attendants noticed that the long naked body stretched out across four free seats under a blanket still wasn't moving. He hadn't taken a meal or a drink, sleeping through the entirety of the ten-hour flight. The attendants in their gold, red and green sarees called on him to wake up; they thought he was drunk, noticing his wet clothes lying in a pile on the floor. For a moment they thought that the naked man was dead. Nobody remembered seeing him board in Mumbai, until a guilty-looking trainee admitted to greeting him, pushed on in a wheelchair by a young Maharashtrian lady, head propped by a flight pillow fitted round his neck, face hidden by an air-pollution mask and sunglasses. He'd sprained his ankle running on the beach in Agonda. He was the allergic type. He was an anxious flyer. There were lots of reasons for tranquillising him and maybe she'd overdone it, explained the young lady. Protocols demanded the hostess wake him to make sure he was oriented, but the lady warned her he was easily agitated; much better if he was left to sleep, for everyone's sake. As she pushed him into the aisle the trainee noticed a wedding ring – beautiful, white patterned gold which matched the one on his finger. No more than twenty years old, she'd assumed the woman was his carer. A call to the desk in Mumbai confirmed the wife had gone back through security before the plane had taken off.

Airport staff brought him from Heathrow to the Unit in the early evening, after a doctor had spoken with the bed manager. It was the last resort as none of the local hospitals would admit him; he was medically stable, too healthy for a bed, and − they didn't have to spell it out − he would be difficult to discharge without a UK address. He was wearing a parka and an old Arsenal tracksuit from airport lost property, but he kept his flip-flops, even on a night in late November, dragging his right leg like an afterthought. The escorts handed over his passport to the admitting nurse: Terence Conrad Davis. Born Lincoln, 8 March 1938. British citizen, the pages full of stamps and visas, mainly Indian and Thai. 'Welcome home Mr Davis,' said the nurse; it was his first time back in the UK for seven years. The face in the photo was clean-shaven, hair worn in a crew cut, a stud in his ear; handsome, impish, on the make, the sort of person who *would* hang out in Asia, tell you *this is the life* with half an eye to scam you; a second property, a new girlfriend. But not this kind of scam, because when the nurse pulled the hood of his parka down that man was no more there than Jesus was in the tomb. Blond greying dreadlocks, long enough to tie in a turban on top of his head, wispy beard, skin a dark, mottled brown leather, scorched by the sun. There was bruising around his sunken eyes, subtle signs of sliding on the right side, more pointedly at the corner of his mouth, which had been covered on the flight by the mask. The scans, though not clear-cut, suggested the main event had happened some time ago, weeks possibly: a stroke in a large cerebral artery. Helping him out of his clothes, they found Sanskrit characters tattooed on the back of his neck, scarring on his scapula and ribs which looked like teethmarks. Well over six foot, he weighed less than eleven stone when he stood in his underwear on the Unit scales, the nurse asking him to confirm his details; a moment passing, and then another; a simple question, but he looked at her as though she was the one unable to make sense. 'On the quiet side ...' the escorts told her.

The Unit's nursing staff were mainly overseas workers from Nigeria, Ghana, the Philippines or the Caribbean. They were braced for late-night arrivals, disoriented, often raucous patients who might mistake you for their mother, brush their teeth with a hairbrush, spit hot tea in your face. But Davis didn't make a sound when they showered him with gloved hands, scoured the filth off his legs, soaked his crotch with iodine – the stroke had left him partly incontinent – or when after roughly towelling him dry two of them held his mouth open and brushed his teeth and cut his long sharp nails lest they scratch the face of an unsuspecting nurse. The dreads would have to go, a risk of hanging. Did he understand … ? Stone blunts scissors; silent as a sculpture as thick coils of rope-like hair fell to the floor. By handover the next morning those who hadn't witnessed the makeover, and met him for the first time seated on the floor of his room, skin the colour of brilliant redwood, silent, pristine, the thinness of the ascetic, the sacred ink on his torso, wrapped in the bed-sheet dhoti, could be forgiven for thinking that this was the first morning in the life of a god.

The brain damage helped hold the fantasy in place. Where many stroked faces are ghoulish, as though cast in the moment of terrible, dawning awareness, his was different; nerve lesions pulling one side upwards, closing the right eye ever so slightly, setting the mouth in a shallow oblique; gently stoic, lightly phlegmatic – '*Really? … Is that what the fates have decided?*' – winking for good measure. A lame right leg, which though it could flex and bear weight, slowed him considerably. Right-handed weakness forced him to use his left so when he ate soup only a fraction of what began on his spoon arrived at his mouth. This semblance of fragility stood out in a place where the threat of assault was persistent; he attracted care, they wanted to help him walk, they wanted to feed him. But he didn't want to eat, he preferred to sit still. Not so fragile or meek, they decided, but a renunciant who has no needs left, whose presence was the food *they* needed, ending their long fast in the desert of the city.

For those first few days every detail of his presentation was organised and then reorganised around someone who wasn't there, like the autumn leaves outside, stirred by the wind to conjure a momentary human shape. His silence reigned, becoming more prophetic the longer it went on. Were he to break it and make a first pronouncement, his carers – many of whom were Evangelicals or Charismatic Catholics – might have imagined he would speak in tongues, noise too apocalyptic for their earthly frames to support. Unless he chose to speak as Adam had to the animals in Eden; one word unique to each of them, a proper name embodying the deepest truth of their nature.

★

On the other side of the city the neuropsychologist's baby daughter, still nameless after thirteen weeks, screamed her way through most of the day and night. He was not immune; the silence did its silent work on him too; Davis was near enough his first ever patient. The holy man appeared to be an easy gig for a newcomer so the Sister made him Davis's 'one-to-one'. This was long before he qualified, beginning his career as a basic rehab assistant acquiring clinical experience. He didn't tell them that before he was a rehab assistant, before he was a father, he'd been a monk.

How the hell had he ended up in a monastery on a Californian clifftop overlooking the edge of the fucking world? He couldn't stand silence. He couldn't bear solitude. Preferring the company of women, he had chosen to sit with dusty old men hunched under yellowing cowls in a damp miserable church, men who broke their silence only to speak a language – *grace, salvation, resurrection, afterlife* – that seemed to exclude him totally. He really hadn't thought this through, fantasising from the mire of his secular life that, God apart, the monk was some sort of archetype, the distillation of all longing, for a mystery that surpasses words; a love affair more pure, more intense, than any he could imagine for himself. Until, that is, he imagined himself into it. *Vanity!* He could not help believe there was a special fate out there waiting for him, which wasn't much less fantastical than the God he couldn't believe in. *All is vanity!* The brothers asked him to choose a new name; Ecclesiastes' *Vanity of vanities, says the preacher.* It was that or Job. He was restless and lonely for most of the three years he was there; for all his longing the language of the world to come had not come any closer.

Near the end of his time he had his one and only 'experience', so near the end, in fact, it's likely he willed it. He had been snaking back along Highway One to Big Sur in an old pick-up, late on a winter afternoon, returning from a shopping trip for the brothers. The ocean was 400 metres below him on one side, on the other a sheer cliff face broken only by steep canyons of redwoods which ran up into the Santa Lucia Range. Despair bore down on him; more lost years, more years watching love die. Something made him decide to leave the road. His foot squeezed the accelerator as he climbed the dirt track that led up to Cone Peak, the message clearer, more urgent every second: the rest of his life depended on him making it to the top of the ridge in time to see the sun set. It came out of nowhere – a rule, the latest injunction of the fates – and he obeyed it. Three years of sitting in the dark chanting ancient songs, the smell of celibate men, their thin warbling voices, ancient words spoken in the dark, the barren deserts of the Near East, stories of suffering and cure, childish supernaturalism, the stupefactions of Rome, none of it had anything to do with him. Until now, flying up the mountain through blazing trees, the sun a darkening clot, an ocean turned to burning oil, those tired old words, those foreign scenes re-formed as urgent promptings; driving as fast as the pick-up would go because he was the possessed young man from Gerasene, he was the legion nameless demons that will not show themselves, he was the swine in search of water to drown themselves, because they like him were cut off, abandoned, insane. That was how he saw it for the briefest of moments; not that he was drowning in his own uselessness in an oceanside retirement facility, tortured by the thought that he hadn't touched a woman in years, but that he had been *excommunicated from the love of the Son of Man*. That's how the thought actually expressed itself in him, *through* him. He must make it to the ridge in time because the Lord performs his exorcisms at sunset. The end of the fucking world was on its way, he feels it, because the real fire, the divining fire is inside, he carries the apocalypse within him.

Then later, sitting on a rock as the stars came out, he was laughing because what he finally understood sounded so mad there was no way of telling anyone. He loves them all – the deaf mutes, the infirm woman, the lepers, the paralytic at Capernaum, the boy with the withered hand, the resurrected dead, all of them were still walking round in all of us, as brilliant as the setting sun. He understood it clearly: the tortured, the demonic were just patients, the patients of his future, calling him; this was the fate, the one true fate, that had waited so long to announce itself. If only he could have made it to the ridge in time, he could have held on to it forever. He missed it, the sea already a dull grey, the forest in shadow, and the feeling drifted. But on that morning with Davis, two years after that afternoon, it began again: Davis had brought him here; speechless, eyes trained on eternity, Davis had called out across the years, *deep calls unto deep*, silence finding itself – across thousands of miles – summoning him from a clifftop monastery in California to a neurobehavioural unit in South London to be his one-to-one.

He was certainly not immune to the supernaturalism he loathed: in the story of one's life there are elements that look like they long to be together. Love, in other words. Within a year he had a partner and a baby, and the baby could not be named.

Davis sits in concentrated prayer, his stare barely reaching a foot into the world; the image of the monk the new rehab assistant had once dreamed of being. He had to pinch himself as a reminder that the man had a brain injury. He had never seen one before the Unit. With beginner's eyes he remembers the sense of a terrible spell being cast, reconfiguring everything – memory, vision, speech, goals, dreams, character – often leaving the rest of the body untouched. One expected the person to return, to heal like other injuries, when instead they drifted further and further away. And once seen, the threat of head injuries popped out everywhere. His daily cycle commute, for example, became hospital bait. A failing front light, the spit of rain, potholes, black ice, oil slicks from leaking cars, their drivers still asleep. No traffic lights at the intersection with Christchurch Avenue, cars flew the junction without stopping. Freewheeling down Salusbury Road in a trance as Kilburn's pound shops and pawnbrokers turned into Infant Cashmere; even at top speed he can't keep up with the tide of gentrification. It was as though nobody was there to make the decision to ignore a light or swerve around broken glass, just as no one decided to make this street rich and not the one before it. Executive functioning (the brain's decision-making suite) part of the capitalist subterfuge: invisible hands, CEOs missing in action. There were head injuries on the Pacific Coastal Highway too; botched suiciders drawn by paradise, unfortunate sightseers who couldn't keep their eyes off it. But they were airbrushed away before he got to see them, like the tourists' trash. Trash was winning here. The yellow siren of a garbage truck flashed – the fifth emergency service – as it hurtled down an empty Westbourne Park Road open-mouthed, and then stopped, suddenly. Techno booming from the cabin as three fluorescent figures with pale amphetamine faces sprung from it. He weaves through them, picking up speed down Bayswater Road, his cadence at the mercy of the tempo of the song in his ear.

Through double iron gates the dark park opened before him; the smell of cold deep air, the memories of the long painful walks he took with Ramona. The path was wide enough for him to close his eyes and feel the rain hardening to hail, hitting his shaven head through the open gills of the helmet, imagining finger-thin wheels on a whitening road. Every ride was ineffably rich with possible injury: others', his own. He could convince himself that in each moment he had the hyperacuity, or the recklessly monumental luck, to ride the city unscathed. Ramona had no such luck. A bearded wire-haired mongrel, she was the first to break his heart. She loved walking so much. Every dog loves walking, and she appeared to love it just as much as every other dog, and given what a walk involved for her – cerebellar damage sustained when she was hit by a car as a puppy had left her with severely impaired motor control in her hind legs – that had to mean she loved it more. The back half of Ramona's body would collapse every few metres, like a pantomime dog where whoever is in charge of the rear is monstrously drunk, creating a terrible torsion on her spine which was enough to send the front half crashing to the ground. Sometimes she would walk tentatively like she was crossing a field of landmines. On other occasions she would forget herself and try to keep up with the back end of a coquettish Dalmatian or give chase to malicious squirrels who toyed with her dreams. However she approached it her neurology would not be cheated; a couple of normal steps then downward dog, creating novel, painful-looking asanas on her way to ground; Splayed Dog, Twisted Dog, Humiliated Dog. Even when she stood still her legs trembled and jigged as though she'd drunk ten pints on the deck of a schooner in a typhoon, meaning that when she ate food or drank water or took a piss – no matter how slowly or carefully – she got covered in food or water or piss.

Sometimes she'd pause in the middle of dinner in a way you never see dogs do, and look balefully towards the bobbing horizon, defeated by the complexity of getting food in her mouth. From the outside, a short walk in the park was a heartbreaking marvel of endurance. From the outside, the basics of independent living were horribly jagged puzzles. From the outside, life was nothing but pain, unstinting shame, endless cruelty. But then what did the outside know? What *could* the outside know about Ramona's inner life, about what it might be like to be her? It was philosophic confusion; the history of her suffering and its indignities piled up in the minds of others making her synonymous with heroic forbearance, stoicism, a lesson to her feckless human counterparts. Really though she just loved walking. Ramona went as berserk with happiness as any other dog at the prospect of a walk, knowing what a walk meant. And it didn't diminish with age, not one bit; that's what he heard anyway, long after he left her to become a monk.

Through the park, the gyratory of Victoria, over the water to Vauxhall. Ramona, Davis, monks, demons, head injuries; love was in the air. Thank God his nameless baby daughter was born intact. The thought of her had also summoned him from monastic life, so that he could in turn summon her into physical reality. The idea of being called by one's future; it made little sense metaphysically, but at the level of experience it was a different matter. He had ridden more carefully in the first weeks after her birth, but there was something about the new job that carried him forward, faster and faster, taking him down a road whitening in the hail, through Camberwell, then out again towards New Cross Gate and a future whose impact was always around the next corner. Risk is a slippery topic in head-injury care. In the years to come a neurosurgical consultant will advise him along with his registrars to go without a helmet, in case it saves them, along with a decompressive craniotomy, from a life of normal span but total dependency. Giving such advice will turn out to be a life-changing mistake for the consultant who will later still be called to a future in which he is successfully sued by the wife of one of his students who took his advice and, following a helmet-less cycling accident and the life-saving surgery performed by said consultant, left her a full-time carer.

Giant off-white houses like Miss Havisham wedding cakes, Deptford's boarded-up pubs, betting shops, reduced by cycling-induced trance to the film-set town: pure focus with zero depth. Only one thought could stop him in his tracks: a future door swung open, out of nowhere, from a stationary car, the driver – celebrating, fucked off, too fat – thoughtless in one of a dozen different ways, timing it to perfection so that he is flung twenty feet in the air, a slow-motion somersault, a London eye, the city seen one last time, a giant spinning heart swelling up to meet his own as, head first, he plummets towards it. Then the long starless night. Dawn happened somewhere over the brow of the hill, a thin grey light from which the silo-like shape of the newly built Unit emerged, in which Davis was silently waiting to be written.

★

The Unit had been built by a private consortium of NHS doctors to capitalise on the shortfall in public provision, their provision. Set in a small industrial estate next to lots for frozen-meat storage, wholesale power tools, specialist martial-arts merchandise, they had invested in a low-slung, flat-roofed version of a future that never arrived. Two businesses were conjoined for economies of scale. To the left, the direction taken by every visiting family, 'The Glade' was dedicated to 'slow-stream rehabilitation'. Even the most zealous families balked at the euphemism, but the phrase allowed the doctors to bid for certain remedial-based funds that wouldn't otherwise be accessible. Slow-stream – the line went – meaning the movement was too slow to see, a canal in fact. What was uncanny was how slowness accelerated ageing – the Unit resembled an old people's home for those in their mid-twenties, forty at most – helped by piped Classic FM, the back catalogue of *Deal or No Deal* on Dave, the piles of *Classic Cars* and *Country Life* magazines (passed on by the doctors themselves, a year or two after publication). With patients that could only do their bidding, the care staff were easy, care-free, jocular, as attentive as one need be given how little was happening.

It was different if one turned right, to 'The Valley', the consortium's neurobehavioural unit. The visitors here were policemen, probation officers, social workers, litigators. To enter meant opening a reinforced, soundproof, seven-digit security door, to see the floor beneath one's feet instantly turn from Carpetright to hard piss-resistant PVC, meet air that was monsoon-heavy with incontinence, Risperidone, puréed dinners. What also struck him – the things that weren't there: door handles, curtain rails, plug points, dressing gowns, taps, fans, detergents, ovens, pens and pencils obviously, cuddly toys; even a 'get well soon' card of sufficient stiffness might, with dedicated application, make it through an artery. The patients were mute most of the time, scattered, only metres apart but with unbridgeable worlds between them. *Patient-centred* drugs had reduced their emotional range to one note, allowing staff the time and space to practise what used to be called 'control and restraint' until it was rebranded 'the management of socially inappropriate behaviour'. On this side the staff existed at low temperatures, spoke in dull, soft voices, part of a general sensory diet designed to reduce by a couple of notches the patients' urge for homicidal explosion. The Unit's two bucolically named wings worked, he thought, as its own naturally occurring experiment; exactly the same staff rotating between the two places, who went from Saga holiday reps to minatory custodians either side of the security door; a totem of English pastoral.

It's hard to remember the shock of being new to it, to recall the strangeness and menace of the place. Nearly twenty-five years ago, thousands of head injuries ago; the inurement was so efficient and effortless. It's hard, but it's crucial that he does it. The women stand out more in his memory, because there were so few of them. Like the lady with Huntington's chorea who lived in the 'Transitional Living Unit' for three years, drinking thickened water to minimise the risk of choking, which took as long as tar to get to her mouth, as though Dali had control of the laws of physics, but only for her. She always wore a brilliant red dress with red shoes, and red lipstick which never held the line of her lips because of her movement disorder. The glamour was uncanny in its effect: still to this day whenever he sees a woman as red as her he fears the colour alone may be enough – in some abstruse way – to make the disease more likely. And the other woman, the civil servant who suffered anoxia giving birth, enough to kill the baby. A park runner who never smoked, overnight cigarettes became the only meaningful unit of her life. Allowed one an hour she'd drag on a Dunhill with such terrible intensity – the deathly inverse of a kiss of life – that it took less than a minute from lighting it to burn her fingers on the stub. And long before it finished she began asking for another, and kept on asking every minute for the remainder of the hour. Thousands and thousands of children incinerated and still she couldn't bring her back. He was a different person then, such sights changed him; it's just he forgot that they had.

Kamal was in the room next to Davis. One of the Unit's first patients (he's in the framed photograph of the opening at reception, being patted on the shoulder by the local MP) he had his own shelf of medical files in the nursing station, eight stuffed tomes, a million words at least:

05.28. K wheels himself in the direction of the sink.

05.33. It has taken five minutes to cover the distance (less than three metres).

05.35. Stationed in front of mirror K removes his vest; his torso is wrapped loosely in toilet roll, greying with dirt.

05.50. K is holding four toothbrushes of different colours in his left hand. He holds a fifth in his right hand which he puts in his mouth and closes his lips around. The brush does not move. His eyes close.

05.53. K takes out the brush and swaps it for the red-coloured brush in his left hand. He puts it in his mouth, closes his eyes and keeps still.

06.16. For the last 26 minutes K has been repeating the toothbrush ritual with each of the five brushes.

06.25. K wraps a fresh roll of toilet paper around his torso.

06.32. K carefully loads each of the brushes with toothpaste.

06.40. He brings each of the toothbrushes to his mouth to taste. The pattern of expression is always the same: surprise followed by extreme disgust.

06.50. K bandages his hands with a fresh roll of toilet paper.

Kamal, Beckett's forgotten masterpiece. Nearly ninety minutes of life taking up one lined page in the medical file; descriptions as transparent and as fine as onion skin and as centre-less. He read through them on lunch breaks, with the sense that, like a librarian in the British Library archives, he was the first and last person to read these words since they were written. The sense too that the recording was as much the pathology as the patient's behaviour; countless words hoarded without argument or purpose, without any flavour of this man's interior, as if transcribing actions are enough to make them legible, add up to a proof that Kamal was in need of the treatment they provided, make them feel that they were still participating in the lives they had fossilised. The net effect was to create a strain of cold, indifferent anonymity. The photo-negative of that beautiful luminous enigma who had just arrived on the ward. Where the silence around Davis was charged with something ineffably precious, Kamal's was ominous, the stillness before a bomb blast. He had been watched more intently than anyone he had ever met. What did that do to a person's soul? According to the files he was labile, cold-blooded, strategic in his violence; that was why he was watched. But flicking through his files it seemed as though Kamal spent all his days wrapping himself in toilet paper or putting different-coloured toothbrushes in his mouth; whatever else the behaviour might mean it was the best way he had found to hide himself from view. The longer the new rehab assistant looked at Kamal – Turkish, a double amputee with powerful arms from all that weight-bearing, but soft, tumescent lips, doe eyes with long lashes, a head that often tilted towards vulnerability, hands with fine, long fingers, ironed trousers with nothing in them – then the more those first impressions consumed themselves, leaving the husk of their opposite; and what he found increasingly difficult not to see was a gentle, maimed, irresistibly feminine creature.

The rehab assistant was new, callow, not yet ground down, still he could see how for the rest of the Unit's carers Davis was an antidote, flying across the world stark naked, called by all that suffering, incontinence, screaming, aggression, attrition, absurdity, to redeem it. Those first few days in his presence were a sort of heaven; nothing had been spoken, and therefore nothing had been corrupted; anything was still possible.What could be more precious than silence in a city? They washed him with choicest soaps brought from home, gently dabbed him dry with their own soft plump towels, massaged coconut oil for his hardened feet, lavender-scented talcum, aloe for the cheeks and forehead that had been cracked by the sun. Men shaved him with a delicacy they hadn't known was theirs. Women brought dishes from home they hadn't made for husbands in years; lion fish, plantain, rare subjee. All because the silence, those fervent eyes, that radiance, allowed a shared, unspoken fantasy dammed by the adversity of working there – the violence, the subduing effort, the daily sight of people deformed so unimaginably – to be acted on. It was nothing, a pause between episodes, a trick of perspective, an accidental posture as his body caught up with the new reality of its injury; but for that moment it felt like they were the ones being looked after.

★

Davis converged with his life in a way he would never forget. This head-injured man so strongly evoked his new baby – his other one-to-one – it could be hard keeping them distinct. Both so shockingly new, silent, barely clothed, changing by the moment, ideas as much as flesh and blood. Their eyes as florid and outlandish as each other's, four of them looking past him like he didn't exist, only where his were seemingly fixed on the divine, hers wouldn't yet stick, had to be bribed to stay ... Who was he? Who would she be? What sort of a name was Terence Davis for something so numinous, so exotic? And what to call her? Still nameless after three months; each one they tried fell off her almost instantly. 'Teflon' they might have called her.

During that brief time she was a portentous infanta: food delivered at the flapping of her crazed fingers, guests shunned, phone calls ended with a look, the squeaking third stair avoided on pain of death. There was something so exact about her needs, so total about her desire for subjugation it circumvented the need for speech. It was pure embodiment; he would measure each of his actions by how it changed the weather in her, first person becoming third, making him her mouthpiece – one moment the ventriloquist, dummy the next. If what the infanta expressed was a language at all it was a private one in which at each new moment the rules changed, and had to be decoded afresh before the world ended, because her scream was so abysmal, and once started could last for days, creating new head injuries – in her, but also in him – with its force.

He came to see her body as an early warning system; eyes were key, then the single deep crease in her forehead, or the tiny spasms in her back at his touch, before they turned into that horrible livid flexing.

Bribery, distraction, cheap magic. He holds up a finger-puppet dog, slowly moving it six inches to the right, the same distance to the left. That way he can see the saccades of her visual system jagging like flies under the control of a trinity of ocular muscles – he had just read about them in his newly bought neuroscience textbook – skipping over the menu of her reality, deciding what to feast on, what to send back. Babies began interested in everything equally, no landscape of preference. By middle age they have nothing but a personalised collection of obsessions – Alban Berg, NA meetings, ear wax, a scrapbook of 'worthy fates', and dogs – already his were creating distortions in her. She tracks the puppet dog's movements drunkenly, at first her gaze a little ahead of it, then stopping though the dog was still moving, then catching up until she's in front again: expectation, regret, exaggeration, obsession. Knowing the ambit of the eye muscles, the names of the cranial nerves that innervate them, freshly learned, he moves the toy out of her field of vision – makes it vanish – forcing her to engage her trunk and lift her head off the corner of the sofa where it was propped – still new, surprisingly heavy, wet-feeling – to continue her attention. The muscles under the milky skin of her abdomen tremble, her neck tautening, the ligaments pulled tight like guy ropes.

'Don't upset her. Where's the sitter?'

'Two minutes away.'

They had a second appointment with a couples therapist.

'We're going to be late.'

The first session had been difficult. She felt that the male therapist preferred his perspective to hers, unconsciously; the therapist's chair was carefully positioned in the middle, making a symmetrical triangle, but the therapist himself always leaned towards her partner. Even when she was talking, the therapist would gaze in his direction, forcing him to compensate by looking directly at her, while she looked imploringly at the therapist: dogs chasing each other's tails. Their failure to talk, explained the therapist, was played out in a baby who screamed so implacably. A scream, he would say, is worth a thousand words; *it implored them to have a conversation, any conversation; it demanded they settle on a name for her; it was the refusal to become an instrument of their relating; or – worst of all – it was a cry to let her be born and to let her return from where she had just come.* (The therapist was called Michael, but he might have been called Ecclesiastes.) The three of them sitting in horrible silence for the last ten minutes of the session, until the therapist said that what they had just experienced had been put in them by her, the scream's negative, its secret source.

His daughter holding the weight of her head up as long as she could, anxious eyes searching, abandoned by her own toy dog, until exhausted the head thuds back into the sofa; her first brain injury. Instantly the eyes bite shut. Watching them in the dining-room mirror, the mother's – freshly fish-tailed with eyeliner, narrow:

'I said don't upset her ...'

Too late. Something unsayable on its way from deep within, a new front of pressure, four limbs pumping in sequence, then a faint sound syncopated with a visible twitch in her diaphragm, and another, a little louder.

'Hiccups ... her first hiccups.'

'She had them last week, and the week before. Pat her tummy.'

The doorbell rings and her little body absorbs the shock.

'That's the sitter ...'

Under the hiccups a sound like a siren at the end of a long tunnel. The puppet returns suddenly, given a squeaky voice – the act is dying – in which it sings like a canary with lungs full of gas.

'Where's the book?'

Another spasm, much larger on the Richter scale. Then like fork lightning the face strikes: a ferocious, vengeful hellhound shocked afresh by each new hiccup, maw at the point of unhinging, back arched to the point of fracture, waiting for the thunder to arrive.

'What did you do to her?'

'Just playing with the—'

Shrieks foul enough to destroy the future.

'She hates that stupid puppet ...'

Pushing her whole young life back at him.

'Where's the fucking book?'

'You had it last.'

HOWL ... HOWL ... HOWL ...

Absolute reproach for all that he had called her to, for this nameless pain that was happening to her nameless blameless self, her screams an affront to the deepest security of them all: that there were always going to be the words to make us feel ok.

★

On the third day

Blessed are the meek, blessed are the sick, blessed are the poor in spirit, blessed are the pure in heart. Davis had brought the monastery to them: stillness, beauty, quiet contemplation. The healthcare assistant was a single mother in her mid–fifties, with grandchildren to look after when she got home from work. Her hair was whitening, turning wispy, her hands clawing, scaled after decades of alcohol rub, cheap emollients, bathing children and adults, but her eyes were still like gemstones waiting to be found, her breasts rode high under her nylon shirt, not yet ready to wilt. She brought her own nut oil for his head, to massage the scalp beneath those dry shorn clumps. What those hands said was private, only for him. He hears them, turning to touch her face, gently tracing an ear with his finger, the lobe thin enough for light to show through. It wasn't allowed but it was happening. Something which could be mistaken for smiling passed across his face, unearthing something in her that hadn't been seen for decades. The pain, the drudgery, the daily assault, was not all there was to see. A smile not entirely his but thoughtless mimicry. And the rest; ruefulness perhaps, anticipating what was about to happen, unless it was like a dog's smile, not there at all, but seen by human need. Hard to know for certain. For there was only one possible meaning when the blessed holy man's erect Shiva linga poked through the makeshift dhoti and still smiling he cupped those age-defying breasts and squeezed the blessed life out of them.

*

Ecclesiastes couldn't stand the silence of his cell in the Hermitage. He had imagined the silence would be spacious, mysterious, enriching, but no, it was terrible, like having your head forced under water, that if he just accepted the new element he might never speak again. So with the monks in lauds or vespers he'd sneak out of the cloister and head for the mountains, or the ocean or the redwood valleys between, followed by Margaret the Virgin. Margaret was not a protecting numen conjured by new-found piety, but a young German shepherd dumped when a puppy outside the church one advent afternoon wrapped in seasonal Walmart swaddling. She was named 'Margaret the Virgin' after the third-century saint from Antioch who was martyred at the age of fifteen. The monks called her 'Marge', but the ironic novice insisted on her full title (having lived so secularly up to this point he wondered – but only to himself – about the authenticity of the saint's title given she died so tragically young, without – so to speak – having fully earned her name; like calling a widower in his eighties a celibate, or an orphan). He met her on his first morning, when as a dashing new monk, half the age of the next youngest, he came across her stretched out lugubriously under the shade of an orange tree in one corner of the cloistered garden. The attraction was mutual; their shining hair, their powerful sinuous limbs, their creatureliness. They saw themselves in each other through young jet eyes, minds never more aware that among the relics they alone were beautiful. *Boundless in their vanity, says the preacher.*

Together they would walk through heavenly landscapes, the last beings on earth, love's pioneers; chancing upon vistas that no couple had shared before, that neither would have ventured to had they been alone; a solitude enhanced. Silences of course, but never awkward ones: he was relieved not to be worrying about what she was thinking, whether she'd had enough to eat, who should pay for what, whether she was cold – even in August. He was relieved to see how nonchalant she seemed about long walks over slippy, muddy ground, about stinging nettles, field mice and spiders. There was relief for her too, he imagined, to be finally close to raw meat, away from those old slow-moving vegetarians who always smelt of broccoli, away too from the sound of the giant monastery bell summoning the Pavlovan monks to church, which still confused her. A partnership forged out of loneliness and misunderstanding, perfected in silence.

Perfect except that Margaret the Virgin had a limp, an accident before he arrived, on the slick kelp beds of Sand Dollar beach, overexcited by her first sight of the sea. A year on and after three surgeries and many hours of physiotherapy, she still wasn't weight-bearing on her front right leg. She compensated, learning to hold the paw inches above the ground, and would prance around the monastic compound like an effete dressage pony. Out in the wilds nothing could stop her chasing deer, wild turkeys, the occasional lynx, through thick forest, or up a craggy mountain, swimming after a stick, leaping after a trout in a mountain lake. He noticed how often, in the heat of the chase she would ground the paw. The vet had always maintained there was no medical explanation for her protecting the leg, no physiological basis for the whelp of agony she expressed when the foot was touched, even at the expectation of it. He thought of Ramona, the dog he had left at a dogs home – Ramona the Orphan – when he had come to the monastery, her courage in the face of real neurological devastation: Margaret the Virgin the Faker, Margaret the Virgin the Malingerer, Margaret the Virgin the Anosognosic (after his neuropsychology training), from the Greek 'not knowing', but carrying the possibility of Freudian denial – about her lack of real injury.

For everything there is a season: a time to embrace and a time to stop embracing, says the preacher. He finds himself snagged by the idea of Margaret the Virgin's co-dependency, its origins, he formulates, in her early abandonment. With some of the continent's most compelling natural habitats on her doorstep, with an unparalleled number of dramatically interesting quarries to pursue, she would just remain in a cage of her own making under the shade of the orange tree if it wasn't for him. It's a *folie à deux*; he feels a terrible pressure to translate the world into zest and wonder for her (Patrick, his future therapist, would hypothesise that he only ever goes out with the dead, tasking himself with bringing them back to life; the same predilection that would lead him to work in NICU), and all he wants is to hide from her in his cell all day. She has placed in him the experience of canine depression. He has to resist it. On their walks he ruminated on how much the monks resembled one another – faces worn smooth with endless repetition – and the wisdom of the adage that people come to resemble their partners and pets and vice versa. Another trick of the mind: he had known all along – lovers, monks, dog-owners, doctors will always choose what they secretly already resemble, unbeknown to them.

She became Olympically fit under his regimen, meaning she required exponentially more and more exercise to be satisfied. He had built his own exquisitely designed torture chamber out of avoidance that turned to love that turned to guilt and then resentment. There was the possibility, still within his reach just about, that all of this was Ecclesiastes' hallucinating, that Margaret the Virgin was being iconically doggish, that her pain was real, that he could only ever see himself in her, would never know anything beyond that. But he *did* know: her behaviour – the head cocking, the spontaneous begging with her one good front paw, the coyote impersonation – felt so artificial, so designed to elicit his favour. *The fate of animals and the fate of man is the same, for all is vanity.* He may still walk her until both of them can walk no more, he may feed her, wash her feet when they get dirty, tend to her if she should ever fall sick (really sick that is) but he will never again tolerate the adoration in Margaret the Virgin's eyes, preferring to look away, turning – like the man in Passerotti's portrait – beyond the frame, towards a future without her, towards the thought of the next one, facing those that judge him from the dark; sad, lost, unable to attach, longing for reality.

*

Dogs were on his mind when the couples therapist suggested they choose a book to read to one another in the evening, as a more involving, more romantic alternative to box sets. On the drive home there was a conversation about which book; the familiar fault lines emerging in the superficial form of a dispute about entertainment vs enrichment, which was really the hackneyed debate about 'what is literature?', which itself was a dispute about who paid the biggest price in being with the other, which was really just another way of not talking. Under that, though, the distant hope that the right choice might just unlock older fantasies, and under that the fear that those fantasies might never work again.

However bleak the therapy session there were always traces of giddiness in its aftermath. *Wuthering Heights* came up: Literature,

Portrait of a Man with a Dog, Bartolomeo Passerotti

check. Entertaining, check. Also tragic, destructive, misogynist, a terrible cliché ... But then only because it's enduring, wild, sublime in its reach, may well be the defibrillator their hearts – flatlined by mundanity, strangled by an implacable unnamed baby – crave.

'What about "Emily"?'

'All your exes are Emily.'

Parking outside the house, they admitted, one after the other, that neither of them had read it in twenty years; its spell still lingered, though.

'"Charlotte"?'

'Puke.'

Then, turning the key of the front door, she looked back and, with the most attractively sheepish look he'd seen in many months, admitted that she didn't think she got all the way through.

'"Branwell"?'

'Get serious.'

He found himself squeezing her from behind, kissing her neck, whispering to her that he's pretty sure he never started it, and if he had, he hadn't got very far in ... The book was already working its magic.

But from the very opening paragraphs his mind was so oriented that it was not love or passion, not the landscape of the moors or the brutality of nineteenth-century agrarian lives, not human caprice and folly, nor the vicissitudes of immortal souls that struck him, but those soul's charges, their wild dependants, the ghastly creatures that snarled and whined at the level of human legs. From its first presentation, the house, Wuthering Heights, was a de facto kennel ruled by a 'huge liver-coloured bitch pointer' and her squealing litter alongside the mongrels, mastiffs and other grim shaggy sheepdogs which haunted its recesses. There was barely any human presence. The creatures themselves were not the doggishly dependable, man's-best-friend variety but 'wretched inmates', 'four-footed fiends', who sneak wolfishly around the place, lips curled, 'white teeth watering for a snatch'.

It went on in the same vein, page after page, the lovers incidental, evicted by rampant, proliferating canine protagonists. He thought the dogs stood for what could not be known, or understood, or loved; the unconscious that is, the dog that each of us carries around inside themselves. They were avatars of their human owners, a little white fluffy thing for the milquetoast Linton, the hyper-sexualised bulldog Skulker for Heathcliff who longs to leap over the explanatory gap and speciesificate. At one point he has to be throttled off Cathy, his 'huge purple tongue hanging half a foot out of his mouth, and his pendent lips streaming with bloody slaver'. How had what was so rabidly pornographic, so canidly brutal, become a Hallmark romantic classic? Reading it to each other night after night, pretending to be moved when all he heard was rutting, frotting, whining; it didn't bring them together so much as remind him about what they hadn't done to one another in many months, about what they couldn't say to one another, about the baby that couldn't be named, that felt like it belonged elsewhere, to a different species. Worst of all he had to bear these perverse associations alone. Until something happened that made him think that she might be seeing things in a similar way, independently of him.

Cleopatra was a brindle-coloured Staffordshire terrier who loved to lie prone on a burnished throne of fake polar-bear skin, front legs supporting her arched back in the regal Sphinx asana, only with dribble leaking from her wide open mouth. Regal in disposition, Cleo was also of the people. She would shamelessly turn on the charisma to secure a stroke, even from strangers, to the point of desperation. Cleopatra, the Queen of Hearts, the People's Princess; recklessly unguarded for a guard dog. The arrival of the new-born baby made her even more fawning as though she considered the unnamed child her own. (They joked about calling the baby Charmion, after Cleopatra's servant, which of course was ludicrous, until they met a Queen's Park mum who had a daughter with that name.) The baby in situ in her cot, Cleo would stretch herself out facing her for hours on end, salivating as usual, and smiling with that wide open mouth, that huge open heart. That's what it looked like. But as the nightly doses of *Wuthering Heights* took hold, that same concentrated stare, the smile, the saliva, turned animal, hungry, jealous, obsessional; grinning ominously because the dog knew what was going to happen. Cleopatra the People's Eater, the Queen of Spades. The thought of what it's like to prise a stick from her jaw when it's locked, how the stick just snaps like a toothpick, a stick twice as thick as a baby's arm. This is how his partner must have imagined it, because fifty pages in and Cleopatra is gone – *please your thoughts with my former fortunes* – a dog-memory, pawed off on her mother's gardener in the Cotswolds. Her only crime: to love too much.

Cleopatra, Ramona, Margaret the Virgin, Heathcliff, Kamal, Davis, an unnamed baby ... watching as everything around him is pulled into a new order. It felt supernatural. The wiser part knew to resist the feeling, that reality was always restless, whorling back and forth, a tornado that sucked up unrelated objects from all over the place – airborne dogs, cyclists with traumatised brains, flying lovers; that couldn't be just stopped and inspected at will, paused while someone tried to turn it into a story. Only life could do that, which never stopped; life the slow artist whose pattern was always still unfolding, always deferring the closing bracket, the meaning reserved for after the last breath.

★

On the fourth day
The door is thicker than normal doors, the wood reinforced with steel bars. There's a slot the size of a letter box five feet from the floor, meaning most of the nurses have to stand on tiptoes or on small makeshift platforms – a diagnostic manual is commonly used – to see through. A shutter is drawn to reveal a slice of room that is ten square metres in total, lit by a harsh strip sunk behind wired glass in the middle of the ceiling. At the same height as the viewer a blue cushioned trim runs round the perimeter, like a high skirting board; below it the sort of buttoned padding one might find on a headboard, to make reading at night more comfortable; above it unadorned concrete painted in a thin 'soothing' yellow. The

floor is cushioned too, like a dojo. In one corner is a squat toilet – the sort found in India – and next to it a sink. Otherwise the room is bare, like a cell in a monastery, only padded, designed to stop head injuries. Sometimes it's 'Solitary' or 'Timeout' but mainly they call it 'The Kennel'.

D is lying on his side, his head propped on an outstretched arm. He writes it down. 'Solitary' apart from the person watching.

D slowly rolls onto his back, shuffling and wriggling, appearing to scratch an itch in this way. The staff are protected. Other patients are protected.

D is pacing from wall to wall. The same rhythm unbroken for forty-five minutes. The patient is in The Kennel for his own benefit.

The lines of his pacing are tighter, as though the walls are closing. He writes it down. The observer is here for the patient's benefit, it's a legal requirement.

He slows, circling around a spot. He sits down. D stares at the grill where the observer watches him. Davis will learn the way a dog will learn, punished because he has not borne the weight of their story.

D makes a howling sound. D bares his teeth at the window. Davis will become what they secretly need, turning into a new story, right in front of them.

He runs towards the door, launching his head at the reinforced glass pane, at him. He leaves a smear of blood on the window. A theatre of pain offered up before a helpless audience.

D is standing inches away from the grill.

He recognises that Davis is punishing him in particular. Only now that he can feel the breath on his face, smell the patient lunch, does the future neuropsychologist recall those terrible forest dogs of Hampi in south India, who like so many fates had followed him here bent on retribution.

That episode began with him waiting on the station platform in Hyderabad for twenty hours as the train was delayed again and again. He knew there was a risk of fog on the line at this time of year, but no reason was given. It was rare for him to move these days. What had begun as a year travelling in Asia after university (ten years before he met Davis) was already two years out and counting, and he'd lost the will to travel. His fellow passengers lay on the filthy floor, their languor perfected with the announcement of each new delay. In London commuters roil at having to wait for five minutes; he can't imagine the consequences of twenty hours at that exchange rate.

It was the middle of the night when the train arrived, half a mile of it slowing, creaking, then dying at exactly the right spot on the platform. He forced his way on, through the bodies hanging on at the carriage door. It was near pitch-black inside, but he knew there were figures inches from his face on every side; spiced breath on his neck, wiry hair against his shoulder, awake at this time because there was nowhere to lie down. Above him children were curled around disused ceiling fans, sleeping somehow. In his first months out here he had taken a berth in Second Class, indulging in the feel of relative wealth: crisp white sheets, air conditioning so powerful he needed a jumper at night, serving staff bringing tea and fried snacks every few hours. These days he travelled General Classification. Even the idea of indulgence seemed quaint and fantastical; there was no possible contrast left inside him where it might be meaningful.

Many hours later the train pulled into Hospet. During the journey light had barely made its way through the thickly grilled windows so it was disconcerting to step outside and find it turning dark. A few tuk-tuk drivers hung out waiting for passengers, hawkers packing up their stalls. He would walk the three miles to the coach station and wait there for the first bus to Hampi and those ancient rocks, formed hundreds of millions of years before in the Dharwar Craton, that he'd been meaning to visit since he first got to India.

There were dogs on the side of the road as he walked. Scratching, stretching, loping. He knew there was a different charge to them at this hour, sloughing off torpor, waking up inside to what was more alive. The light was failing and the battery on his phone had gone, his only torch. The houses and storefronts disappeared. Soon the road was lined only with jungle. There were no cars at this hour, only the occasional goods truck bound for Bangalore. The clouds were heavy-looking in the sky, barely above the treetops. It could rain, even at this time of year. He knew they would follow him, that disinterested, disparate they would be stirred by his passing and gather in his wake. It was a knack. It was a curse. More appeared from the trees, as though summoned; a stocky male, two females, their dugs swinging, smaller shaggy Tibetan types – there was a monastery nearby – a mastiff cross and two of her pups, others that were hard to describe other than their doggishness. They packed around him in a way he had not seen before: four or five out front, more flanking him on either side, but most behind, panting, growling, at his heel. So when they left the road to follow a single track which headed into the jungle, he followed, moving at their pace.

Slowing, he searched in his bag, remembering a packet of biscuits; it's possible they smelled them. There was a sharp nip on the back of his leg, not enough to draw blood. A low growl built between them, but he wouldn't be cowed. He opened the packet with his teeth and scattered the biscuits around him in a circle. Instantly he regretted it; they swarmed, mauling one another terribly to get at the crumbs. But moments later the biscuits had gone, conflicts forgotten and they circled again, snarling in unison, facing him, choreographed by the same ominous urge, so that when he tried to take another step forward there was frenzy. The time for walking had passed. But being still, holding out calming hands, gently offering soothing sounds also made it worse. Fifty dogs or more, far more than he's ever seen before together, syncopating each other's aggression until it was a near continuo, which is somehow conducted so that a momentary lull triggers another round, more intent than the last. And yet there was no conductor, nothing that's obviously caused it, nothing to decide to keep it going. Each one of them was, like himself, helpless. Maybe it will go on forever like this.

For no reason the barking stopped suddenly. The silence that followed was another pact but he is part of it this time, a stakeholder rather than the victim; or that's what he had decided – it was just a feeling, but it seems to hold. He sat down carefully. Still they were pacing, agitating around him, but separate from each other again. He had formed a contract; if he kept still, if he remained silent, they would settle. Though they were in a clearing in the jungle, the low clouds overhead made it especially dark. The dogs' snouts were close enough to touch him, for him to feel the heat of their breath, drops of saliva falling on the ground by his feet. Every now and then the barking began again, deafeningly loud, out of nowhere, propelling itself for another round, lasting as long as the sound alone dictated. But he had stopped taking it personally, he had stopped telling a story about it: this had nothing to do with him, nothing to do with any of them. The pack decides, for as long as he submitted to it he was one of them. He closed his eyes and in the boundary-less space beyond his eyelids he can see the shapes they make, heads bent into crotches, backs arching, legs stretched out before them.

In this way they spent the night, the dogs and he, packed together in the forest; silence and stillness, insuring each other. The dawn light cracked open his eyes. They are not there.

'Bhu-bhu-bho-bho.'

Like a bark.

'Bhu-bhu-bho-bho.'

Davis's first words . . .

'Bhu-bhu-bho-bho.'

If a dog could speak, still we could not understand him.

'Bhu-bhu-bho-bho. Bhu-bhu-bho-bho.'

This is what Davis has been waiting to say, this is what lay on the other side of that silence. Not enlightenment, not wisdom, but aphasia and animal noises. Nothing could have deconstructed him so efficiently, or re-embody him more startlingly and with such speed than utterances that fit so perfectly with who he has become, which is also what they have turned him into.

He runs towards the door again head-first: Davis is giving himself a head injury in a head-injury unit, like the possessed Gadarene swine exorcising itself. The observer can't bear to watch; he has to watch as the protagonist headbutts the fourth wall.

Davis retreats out of view and whines: 'Bhu-bhu-bho-bho. Bhu-bhu-bho-bho.' Then silence. He is there and not there, Schrödinger's dog. Really they are in solitary confinement together: Davis giving him an experience of what it's like to be him, of the pain he's in. Now it's his pain too; the pain of helpless spectation, the pain of complicity. He looks away to see Kamal there in his wheelchair looking at him wistfully. He looks back at Davis who is winking, then to Kamal who is also winking; an actor and his audience and *his* audience. Only it's tragedy not comedy they are watching, a tragedy because no real interaction is possible, because the audience can never stretch out across the dark theatre and comfort a character. Unless, that is, he becomes a character himself. But that wouldn't happen until years later.

★

On the fifth day

Two detectives turned up dressed in bomber jackets, designer jeans, running shoes, like boutique shoplifters. Though young, they had the glaze off-pat, as though whatever the future turns out to be, they have seen it before, for a living. The speech and language therapist gave them a lesson in the disorienting effects of haemorrhagic strokes, the potential devastation of cognition, the neuropsychiatric sequelae, the transformation of personality. They nodded and chewed their gum.

'This patient has a complex language disorder that severely limits his communication, meaning he wouldn't be able to answer questions without a lot of support.'

As the rehab assistant listened to her from the doorway (he'd been Davis's one-to-one so long nobody noticed he was there anymore, like he'd been absorbed by his patient) he couldn't help think of how quickly and confidently the voice of reason had returned, camouflaging the collective insanity of the last few days.

'Even though he looks like he's defiant or aggressive, sometimes it's just a look.'

The detectives understand.

'He can't help it.'

Thank you for the background.

'All I'm saying is that whatever he says or appears to acknowledge, it can't be relied on at the moment.'

That's very helpful.

But they don't look like they've been helped.

Seated round a table in a small consulting room they showed Davis a photograph: a woman in her early twenties from Pune in west India, young enough to be his daughter, the wife who left him on the plane. His eyes soften, prick with tears.

'Bhu-bhu-bho-bho.' Plaintively.

'Bhu-bhu-bho-bho.' Longingly.

(The rehab assistant transcribes it this way – the Marathi for 'woof woof' but really there were different rhythms, stresses and pitches, human language too coarse to do them justice.)

'Bhu-bhu-bho-bho.'

They wonder why would she dump him on a plane rather than look after him?

'Bhu-bhu-bho-bho.' Heartbroken.

Because she didn't want to be saddled, not after all she'd been through.

'Bhu-bhu-bho-bho.' Defiant.

'Bhu–bhu–bho–bho.' Cracking.

The speech and language therapist tells them this type of unsupported conversation is upsetting for him.

'Bhu–bhu–bho–bho.'

She's already explained to them; he can't understand their questions.

'Bhu–bhu–bho–bho.'

Let alone answer them.

'Bhu–bhu–bho–bho.'

But they've heard that before.

'Bhu–bhu–bho–bho.'

And they've seen this before.

'Bhu–bhu–bho–bho.'

'I'm no expert but I think he gets it,' says the detective.

'Bhu–bhu–bho–bho.' Melancholically.

'We spoke to her, Terry.'

'Bhu–bhu–bho–bho.' More melancholically.

'Tell us, what did you do to her?'

'Bhu–bhu–bho–bho—' Urgently.

The speech and language therapist calls a halt. 'It's too distressing.'

They tell her it's supposed to be distressing, he's done distressing things.

'Bhu–bhu–bho–bho.'

That's how they work it.

'Bhu–bhu–bho–bho.'

They might not be speech specialists, but they know guilt when they hear it.

'Bhu–bhu–bho–bho.'

'Why haven't you been home in seven years?'

'Bhu–bhu–bho–bho.'

'Drop the act Terry.'

'Bhu–bhu–bho–bho—'

'What are you hiding from?'

'Bhu-bhu-bho-bho   ...   Bhu-bhu-bho-bho   ...
Bhu-bhu-bho-bho.'

'Alright, have it your way Scooby Doo ...'

He'll change his tune. They've seen it hundreds of times
before. Makes no odds to them. They can come back; a ride out
of the station, a fish curry at the Sri Lankan nearby afterwards.

★

On the sixth day

Another chapter begins: first swami, then predator, then
mongrel, cur, a demi-wolf that preys on women, young and
old: same information – silence, staring, starving, bite marks,
smiling – reconfigured in the opposite direction, to meet
another need, no more conscious, no less desperate.

It took the Unit's neuropsychologist to stop the madness.
She was a short South African woman in her early thirties,
sharp-featured, a small hunch on her back. It looked like a
foetus had crawled around its mother and clambered
upwards, resting where the shoulder blade was. (That's what
it reminded him of; his partner pregnant again, accidentally,
the timing couldn't have been worse.) She wasn't there for
long, a short working visa, only a few weeks, just enough
to change the course of his life. She asked for all the staff
to join her in the teaching room, like an old-school Grand
Round, where she sat at the front table, a cushion on her
seat to raise her a little, Davis sitting next to her, so obedient-
looking, so proud and serious, knowing he was the guest
of honour; you might as well have pinned the rosette on
him there and then. She produced a biscuit and held it
before him, a married couple performing their favourite
dinner-party skit.

'Would you like a cookie Terry?'

He snatched it out of her hand and stuffed it in his mouth
before she had finished her sentence. 'Bhu-bhu-bho-bho.'

'You all know Terry has powerfully impulsive behaviour, caused by disruption to frontal areas. Would you like another one?' She holds one up. 'Likely, he's someone who's always had big appetites.' Like an impresario she turns to her volunteer, who is hypnotised by the biscuit. 'If you're able to wait while I count to three – I'll give you two for the price of one.' She points to two biscuits on the table.

'Bhu–bhu–bho–bho.' Eagerly.

'One ... two ...'

He snatches it out of her hand and stuffs it in his mouth again. Turning back to her audience: 'He can't wait, or think strategically. Or if he can then the thinking can't withstand the urge – that limbic, animal urge. A child of six can usually restrain herself in such situations, a child of four can't wait.' There was a precision to the way she talked, confident that she could bring all this wildness under control, just by naming it correctly. *Like Adam.* The right words to say; this more than anything else is what the situation needed. She made a short speech about how the brain had evolved alongside the world but at a different pace – palaeolithic longing, classical refinement, medieval institutions, Internet porn housed next to each other forming an unwieldy transhistorical compound. (She didn't say it quite like that, but that's how he would remember it.) How could we expect each other to bring coherence to all this, even the intact ones?

'It's not just food, other sensory stimuli elicit similar impulsivity,' holding up a biscuit in one hand, a piece of bubble wrap in the other. Davis snatches the wrap out of her hands, then the biscuit, and sticks them both in his mouth.

'The stuff of the world is just so put-in-your-mouthable ... Leaving aside how much he understands instructions for a moment, he's hopelessly stimulus-bound, like a baby he can't delay his wants, zero executive control.'

The bubble wrap pops in Davis's mouth until she holds her hand out and he drops it into it obediently.

'Crucially, the more of it there is, the less able he is to resist it.'

He was no different from anyone else, just unable to conceal it. The diminutive maestro places a satsuma before him, then an orange, and then a large grapefruit. Before Davis has time to consider the options the grapefruit is in his hands and he's sinking teeth into its peel.

'It's the same with bits of the anatomy. Bigger is just more grabbable. Really we have no idea about his sexual proclivities. Yes he was married to a much younger woman, but he's not the only one. And yes he's priapic on occasion because the stroke has affected his thalamus, not necessarily because he likes the look of what's in front of him.'

Twitters of disbelief. The neuropsychologist recommends female staff wear baggy tops over their t-shirts or loose aprons when they're working with him. She recommends that Davis is fed a sensory diet, given tactile and olfactory stimulation as a way of capping his arousal at a tolerable level. Most important she suggests that they, his carers, find a way of keeping the right distance, not too close, nor too far away; that's their job. She stops talking for a moment. They know what she is referring to: their undigested urges, their lonely hearts; not confusing those with him.

'What is your date of birth, Terry?'

'Bhu-bhu-bho-bho.'

'What is your mother's maiden name?'

'Bhu-bhu-bho-bho.'

'Classic Broca's aphasia, stuttering monosyllabic plosives. But you also notice how he strings the syllables into different lengths, gives them different metre and prosody like a baby's babble. Listen again. Terry, what is the capital of Nicaragua?'

'Bhu-bhu-bho-bho.' Nonplussed.

'It's Managua. Compare that response to: what are your children's names?'

'Bhu-bhu-bho-bho ... Bhu-bhu-bho-bho!' Wild remonstrances.

'I know ... I know,' she calms him before turning back to her audience. 'You see the difference of course. Do you think he understands? I think he understands. But do you think he thinks I understand his response? You have two girls don't you, Terry?'

'Bhu–bhu–bho–bho.'

'You see he believes he's answering meaningfully. He believes I hear his daughters' names, even though all I really hear are the meaningless syllables. And yet when I repeat them back to him' – turning to Davis – 'Bhu–bhu–bho–bho and Bhu–bhu–bho–bho?'

Davis cocks his head, makes a puzzled face.

'All he hears is my nonsense. Bhu–bhu–bho–bho and Bhu–bhu–bho–bho?'

He is shaking his head violently: 'Bhu–bhu–bho–bho ... Bhu–bhu–bho–bho ...'

'Comprehension is intact, it's the feedback and expression that are destroyed, either because he can't hear what he's actually saying – there's no evidence for auditory impairment – or some of his circuitry is giving him erroneous information about what emerges from his mouth.'

'Bhu–bhu–bho–bho ... Bhu–bhu–bho–bho ...' Proudly restating the names of his children.

'Like a baby, he means what he says and the world is expected to do what it's told, however it's told. As far as he's concerned we are the dumb ones.'

'Bhu–bhu–bho–bho ... Bhu–bhu–bho–bho ...' Terry continues.

'He has anosognosia for his own aphasia. And we have anosognosia for his intentions. Not to mention anosognosia for our own. We are all in it together.'

'Bhu–bhu–bho–bho ... Bhu–bhu–bho–bho ...'

'Can you imagine what it's like to be him, how wild that must be?'

'Bhu–bhu–bho–bho ... Bhu–bhu–bho–bho ...'

'To wake up tomorrow and, without your knowing, the entire world has switched languages leaving you as the only native speaker. Can you ... ?'

'Bhu–bhu–bho–bho ... Bhu–bhu–bho–bho ...'

'Terry-speak, a language of one.'

She let silence take hold again, seeming so sure of what she was saying, how she was saying it.

'Of course you can't, you can only imagine yourselves in that position, and that has nothing to do with him at all.'

All was vanity, all was shepherding the wind. For a moment she had brought it all together, like an artist shepherding the wind. She died a few years later in her sleep from a short respiratory illness. Davis would most likely be dead by now too. The encounter with her, with both of them, affected him. He could see how one person thinking straight could alter the momentum of reality, save a patient from the fate of other people's minds. As fates go it was as humble as it was miraculous. That's when he made the decision to train as a neuropsychologist himself, right there in that room at that moment, twenty-five years ago.

'And what about his wife?' someone asked.

'Which one?' she replied. The end of one story and the beginning of another. The police had found out that the young Indian woman who dumped him on a plane was not his only wife. He had another whom they'd managed to track down. Neither she nor her two grown-up children had heard from Davis in seven years.

'Nothing I've just said doesn't mean Terry wasn't a bigamist or a charlatan with aggressive tendencies before he had his stroke, doesn't mean that he didn't always take more than he should have, want two for the price of one.'

★

On the seventh day

It was unusual to see a well-dressed middle-aged woman who was not a lawyer asking for a patient at The Valley reception. She'd travelled down on the train from Morley, just outside Leeds. She looked older than him, but still she was handsome, and proud: skinny jeans with high heels, a pale cream silk shirt under a cashmere coat. The staff caught her up with the facts, prepared her as best they could for what he'd be like, seven years, one wife and a devastating stroke later. Then they took her to him.

Standing before one another in silence, their faces accounting the different lives they'd had; she seemed to grow older but only because he turned younger, right before her, more familiar, more boyish, not quite able to look her in the eye.

'Bhu-bhu-bho-bho. Bhu-bhu-bho-bho.'

'It's all he can say,' the rehab assistant explains.

'Bhu-bhu-bho-bho.' Plaintively.

'Bhu-bhu-bho-bho.' Regretfully.

'Bhu-bhu-bho-bho.' Tenderly.

'... It's nonsense but he doesn't know that,' the teaching had survived that far.

'Bhu-bhu-bho-bho ... Bhu-bhu-bho-bho.'

'Careful, he can be aggressive ...'

'Bhu-bhu-bho-bho ... Bhu-bhu-bho-bho.'

The assistant sees her looking at the beautiful white gold patterned ring loose on Davis's finger, noticing the thin coppery band around her own.

'Bhu-bhu-bho-bho.'

She doesn't seem remotely surprised – taking her coat off, making herself a cup of tea – instantly familiar, quickly losing interest, as though this is how he's always been.

'Bhu-bhu-bho-bho ... Bhu-bhu-bho-bho ... Bhu-bhu-bho-bho.'

... banging on about the same thing over and over: himself in other words. She will tell the rehab assistant how their life together was a pile-up of scenes like this one, standing in front of each other, catching up with what can't be said: the first kiss outside Wigton Moor Youth Club disco; the bank holiday ride on his Enfield in the pouring rain through the Dales which ended with him proposing; the morning she told him she was pregnant and watched him catch his breath – she'd bring in photos taken around the same time – and the many occasions he promised her they'd 'bugger off' – Thailand, India, wherever she wanted, soon as the last kid left home – the beginning of something totally new, he promised her.

Looking at the photos – together on honeymoon with his Teddy boy quiff; in the back garden with the baby daughters, joking with friends in a beer garden – he is struck by what a dandy their holy man had always been, a born show-off: messing around, but always half an eye looking out of the frame, and half of that half not joking but lost. Beer gardens, northern council estates, package holidays, in beautifully tailored shirts, pomaded hair, two-tone Italian leather shoes: this is where he had really come from, this is who he really is, he thinks. *All is vanity.* Would his colleagues feel as sheepish as he does seeing this charlatan, this wide boy, then remembering the unadorned god of the first days, the whole Unit clambering over itself to touch the hem of his garment?

'Bhu–bhu–bho–bho … Bhu–bhu–bho–bho.'

'Never could stop talking about himself.'

'Bhu–bhu–bho–bho … Bhu–bhu–bho–bho.'

There is nothing new under the sun, says the preacher.

'Bhu–bhu–bho–bho … Bhu–bhu–bho–bho.'

Just more of the same.

'Bhu–bhu–bho–bho … Bhu–bhu–bho–bho.'

She left without saying goodbye. Davis waited by the button-pad door, all afternoon and into the evening.

★

After three years in the company of old men and a lame, codependent dog he was desperate to fall in love again and for that love to make him a father. For the briefest of moments there was the sense that anything was possible; timeless light-filled mornings on the heath, no need to talk because everything is understood, or a language of their own. A little over a year later, they find themselves at the impossible place; no way of talking, no dog to walk, longing undimmed, an implacable child, still nameless, and now another on the way. Whoever picked her up was holding a hand grenade without a pin, the screaming, night after night after night. Not hunger or incontinence or any mere physical pain could cause this degree, this timbre of crying; it had to be a more primal trauma, some terrible verdict on their coming together.

'Bhu-bhu-bho-bho,' whispering it softly. 'Bhu-bhu-bho-bho.'

'...'

'Bhu-bhu-bho-bho.'

'...'

'Bhu-bhu-bho-bho, bhu-bhu-bho-bho, bhu-bhu-bho-bho, bhu-bhu-bho-bho,' over and over and over, 'bhu-bhu-bho-bho, bhu-bhu-bho-bho,' until the muscles in her forehead loosen, the arch of her eyebrows relaxes, her cheeks go slack. 'Bhu-bhu-bho-bho, bhu-bhu-bho-bho,' convincing himself she knows what it means.

'B-b-b-b-boo,' gurgling back at him.

'Yes ... Bhu-bhu-bho-bho,' encouragingly.

Her mother appears at his side in time to see another fantasy coming to life.

'Bhu-bhu-bho-bho.'

'B-b-b-b-boo.' Pleased with herself.

In unison, 'Bhu-bhu-bho-bho.' The end of the dogfight, mother and father together at last. 'Bhu-bhu-bho-bho.'

'B-b-b-b-bro.' The sound of hearts thawing.

'Bronny? She just said "Bronny".'

'I don't think so.'

'You're so cynical ... Clever girl. Bronny ... Bronwen.'

'Seriously?'

'Definitely, Bronwen.'

'Bronwen, then. Bronwen the Virgin.'

★

It sounds like a just-so story in the retelling, but at the time it was nothing of the sort. It was the coming together of a history, their history, in all of its reality across the course of a week or so, and it changed everything that came after it. It was perfectly normal, without a shred of mystery. Because family life, monastic life, hospital life, canine life – they were always theatres of the everyday, and as some philosopher once said, there is no theatre that is not prophecy; that autumn must follow summer and spring winter, that there are four elements, that there is happiness, that there are innumerable miseries, that life is a reality, that life is a dream, that man lives in peace, that man lives by blood, that the worst has not happened, is always on its way. A few weeks later Mrs Davis came back and stayed overnight at a nearby Holiday Inn. She'd brought Davis a few old shirts that she hadn't thrown or given away, and some motorcycle magazines that he used to read. She sat in the dining area talking to some of the other patients. Kamal let her feed him. They found each other funny for reasons nobody else could understand. While she was there Davis followed her wherever she went. She didn't pay him much attention, but it didn't stop him. He'd sit near where she sat, as quiet as he was when he first arrived, only now nobody noticed. She would keep coming back, once a month, doing some shopping in town beforehand and having dinner in the Sri Lankan restaurant after her visit. The whole Unit looked forward to her coming, a new antidote to replace the old one, bringing as she did a fresh kind of spirit to the place which was bright, light-hearted, a little glamorous even. She always enjoyed combing the Huntington lady's hair and fixing her wobbly lipstick. Then she'd feed Kamal, making sure her husband could see her while she did.

Jim

'Start with you standing on the street corner, in front of the Korean. It's hotter than it's ever been. You feel the heat wanting to bend you like an iron bar. You inhale warm sweet meat, petrol, sweat mixed with deodorant. You hear conversations in English, Arabic, Marathi, a barking dog, children screaming; under it all the sound of your own breath. Now look up at the sky. For once it's not the same colour as the road; a memorable blue, a high-ceilinged Tibetan blue. You see brilliant lime parakeets fizzing low over the river. Look left towards the Broadway, traffic still as a photograph, then right down Putney Bridge Road, shimmering in the heat, your future receding south over the Thames. Below you see rowers holding a line that keeps them in the shade of the trees. Details will be critical to you Jim; like the willow branches trailing in dark, fast-flowing water, the moment one of them hooks a floating crisp packet. People have left work early, like buffalo the heat makes them head for the river. Standing outside the Blue Anchor holding cigarettes and drinks in the same hand, using the spare one to make themselves understood; expansive once more. Study them, notice how each inaudible face lands differently on your nervous system: regret, urgency, boredom, ecstasy. A moment's thought, after which you know them better than they know themselves.

'Turn the spotlight on your body for a moment. Feel the parts of you sticking to your clothes, feel other parts of you touching yourself: a fold in your stomach, the bases of your fingers and toes. If it was the skin of another person up against yours, you would feel it – more than feel it, you would read into it. Fight against the dullness, bring your body back to life, because sensitivity is all that can save you. Then feel the current of warm air moving over your knuckles when you scratch your nose, the tautness of the different muscles in your forehead, the dry creak of your eyes each time they move, most of all the innuendo of unfelt threat that is keeping your jaw locked tight. Then relax, let yourself go entirely, keep going ... a little further ... the fight that's about to happen has already taken place. Keep noticing.'

Jim is not there. Because if noticing takes place in a small room inside us, behind the glass of an observation station somewhere behind the eyes, then where is the other observation station from which we notice that one? Really there's just an ineffable, self-less plane where all noticing takes place, a field without limit, noticing without a subject. Jim is not there. But to have any chance of recovering he will need to be there again.

'Now you're back in the taxi, your beautiful black cab, a second home paid off in full. The traffic is finally breaking up. You're driving over the bridge, Drake's "In My Feelings" on Capital. Notice how music affects the way you see people, overlays their movements; makes walking walking, children more ecstatic, strangers stranger – bringing another life to their faces, one that's not entirely their own. It's all part of your training. You see white flesh squeezed into pink polo shirts, eyeless behind fake Ray-Bans. They've waited all through winter for this, to spring these outfits from their wardrobe; the first day of endless summer. Hotter than Africa, unforgettably hot. Your fingers rap the steering wheel to the beat, it relaxes you, and makes the passengers feel reassured; it says this car still has a driver, he is part of the same reality as you. But now there are no passengers. A shaft of bright silver light bounces off the mirror, making pedestrians catch fire. The world heated past the point of no return. You're heading straight for the sun. There is nothing left to be done. You pull the visor down. You put your sunglasses on. You drive to the music.

'Roadworks on the South Circular. Through windscreens unplugged faces the colour of kerbstones. The traffic hasn't moved for ten minutes. You know there's no way round it. You're an expert, 50,000 hours on the meter and counting; Kilburn Lane, Seymour Place, The Cut, Fournier Street, the Isle of Dogs; millions of addresses stitched into the soft white machine nested in your head. Now though you must remember something new. Now you must rediscover a lost scene in an imaginary location. Now you must become a different kind of expert, in suffering. Other eternity in traffic: can't move, can't move on.

'A group of men spill out of the White Bear, loose ties, slack-looking faces, unpredictable movements. The smell of cocaine and sweet designer cider. Their laughter has teeth. Your hands whiten around the steering wheel, jaw cemented by the gum in your mouth, guts thick as rope. Intuitions you call them, but really you're just catching up with the body's instant prophecy, which is really an ancient memory. There's a chance they will want to go to Brighton on a whim. Without a thought you switch the sign off and close the windows. "Brighton? It's not worth it." If it was a thought, it's an after-thought, the action already done. One of them makes the wanker sign at you. "Is that really how he wanks?" you think. So casual, so half-hearted. They move on out of view. The traffic lights change to green but there is no space to go forward.

'The breeze blocks stacked outside your house mean nobody can take your space. You park up, the rest of the journey one of life's trillion absences; you don't have to remember everything. You walk up the path to the front door. Screaming kids, dogs barking; they mean nothing yet. Easy does it. You're whistling. There's a spring in your step. You say hello to Frank next door. (You did, he remembers it: "a spring in your step" are his words.) You decide you'll go fishing early in the morning. (You love fishing, it's a fact; laminated on a card above your hospital bed, along with pictures of your family, because even facts, even the things you love have to be stored outside of you now.) You walk up the path. The back door is open. You shout Carol's name. You will go fishing early and be back in time to take them out for the day. Still whistling "In My Feelings", you put the takeaway down on the kitchen table. Nobody answers. Don't panic, she must have told you she was popping out when she phoned you earlier with the order. Remember? Don't panic, there's nothing wrong yet. You just forgot, when forgetting was still normal.

'But there's an important principle to get straight; for this to work we must have a threshold separating before your injury from after, like a hospital door. Easier said than done: the closer you get to the door, the more trauma plays games, tricks you into thinking you're already on the other side. Don't be fooled. Remain innocent for as long as you can: just dogs barking, children screaming, an empty house. She is not here because she has just popped out like she told you and you forgot. Let's go back and insert what you forgot, it'll only take a moment. You are driving over the bridge. You are heading straight home. The light is bouncing off the river. You put your glasses on. You call Carol. She asks you to pick up something for dinner. She tells you she is just popping out to see a friend, she'll be back soon. You relax, like you've already gone fishing; the stillness of the river at that time, as deep as England.'

The neuropsychologist met Carol and the girls once, the only time they came to visit. They were dressed for a family do, their faces serious under thick make-up, as though they couldn't quite trust what would come out of Dad's mouth next. That's the sense the neuropsychologist had anyway. They looked like they wanted to jump on his lap and pull on his hair, but the door couldn't be gone through in the other direction.

'The house is empty when you get there. You lay the table. You think of the time a few summers ago when you piled everyone into the back of the cab and headed to Broadstairs. A weekend in Broadstairs. Hannah and Miley were toddlers. They played with starfish in rock pools until it turned dark while Carol read and you fished from the beach. And when you'd had enough of not catching anything, you watched them – drinking a beer from a nearby rock – as they fed stars made of sand to starfish, while the first stars came up over an emptying beach. As you drive back west into London, the sun is going down on the longest day of their young lives. Catch it Daddy, faster, as it disappears behind city buildings, faster, before the world ends. They fell asleep suddenly, heads thrown back, mouths wide open, like dead fish. That's what memory will feel like to you: a colour, a song, yours again.'

But really belonging to someone else, the neuropsychologist himself in this instance; a day at the beach with his kids after the split, clearing his mind of the patient on NICU.

'You enjoy taking care over the spread, you have a soft side that most don't get to see. More screaming. You are not concerned; instantly you know it's pleasure you hear, the yelp of summer. You're at home, in your skin, relaxed, still on the other side of the door. There'll be a point in about six minutes, when you'll be more terrified than you've ever been, but I'll let you know when that's coming, so stay as relaxed as you can for now. You could sit down and wait for them with the paper, they won't be long. Or why not go out and find them?

They won't have gone far, and it's such a beautiful warm day, same temperature as Riyadh.

'The estate looks softer in this weather. You turn up one street, down another, past the launderette, the Tanz-in-'ere Solarium, rows of empty green and brown bins, maisonettes with half-complete extensions, St George flags draped from windows ready for a new crusade. You turn left and then right, and then left again at the postbox; moving steadily towards your destination like the pulsing blue dot on the phone maps. You aren't allowed to use those maps; your memory is your reputation, a high-wire act without a net; it's what makes you expensive. Down ginnels lacquered with "anti-climb paint", past an electricity substation with empty shoes on the wrong side of the high metal fence, across a car park where nobody dares park. You cross a footbridge which spans a railway track, back into more of the same houses, roads with colonial names. Even this far from home you know them all, their low frequency exoticism made them easier to learn. Children are water-fighting with hoses. Dogs are barking. You walk through a foul-smelling subway under the dual carriageway. You know the way without thinking. You have the knowledge. You will remember it again. Turn left, climb the stairs, go through the green iron gate. You have reached your destination,' says the neuropsychologist, feeling like a stand-up addressing a darkened room, in the hope that somebody's out there listening. The directions would be remembered, he thinks, because they are stored somewhere other than Jim's Rolls-Royce of a hippocampus, deep in the ancient brain. This is what he's mining for, drilling the same tale into him again and again, bypassing the part of Jim's brain that's gone (the Roller nicked in broad daylight by a bunch of kids, right before his eyes).

'There must be 300, maybe 400 kids of different shapes and sizes on the common, evicted for the afternoon. Yours are here but you can't see them. The detail is so overloading, the reality so perfectly rendered, it becomes a painting: a

Bruegel. You pan across slowly from right to left, beginning at the toddlers' playground. The climbing frames threaded with twisted torsos, severed heads, amputated limbs – more Bosch than Bruegel – the best of surgeons can't stitch these together, and yet your brain can do it in an instant, discerning that none of them belong to you. You see a screeching multi-corpse pile-up of boys at the bottom of the slide; younger girls quietly, thoughtfully building pyramids out of sand misshapes in the dog-fouling area; a dozen kids perched on top of the swing frame, fat as pigeons. Across from the playground, a large open field hosts a football game, forty-a-side, clothes for goalposts, a line-less pitch stretching in every direction depending on where the action strays. You can't see the ball, just midgets making patterns, shared vectors of running. Even though the detail is inordinate – the clothes, the distant features, the way they shout GOALLLLLLL, the attitude of their bodies in space – the calculus is performed in a fraction of a second: these kids have nothing to do with you. Slaloming through the pitch, as though overlaid in a different story, smaller children ride on bikes, chased by a dozen more dogs, cross-hatching a chain of girls holding hands to keep together, singing Drake – he was contagious that summer – in a round. Where the grass turns to scrub you see older boys standing in sprigs, side by side, talking to each other with their phones. At the picture's edge, another high metal fence topped with barbed wire marking the boundary of a large industrial estate, and two chestnut dray horses, facing in different directions, unaware of the other, the only stillness left in the world.

'Peek through the door Jim, go on, just a quickie. See how everything flips the moment you're on the other side: suddenly nothing is innocent, everything is charged with menace – the darkening sky, crows dotted on the grass, smudged lipstick, laughter curdling, and your children's faces in every face you meet ... See what happened there? You glimpsed something before you'd arrived. The future threaded your eye and pulled you through. Take a breath. Horrible wasn't it? You have to believe there was a time before the trauma happened, take it on trust, a faith of sorts. Without it you are lost forever. We are still at a point when the threat you feel is nothing more than the shadow that follows all fathers: concern for their kids. You must try to remain that way for the next two minutes or so; innocent like kids, so that when we turn the page and the real threat arrives it can pull out its claw, suddenly, from deep inside its long overcoat, on the unsuspecting audience. Until then, soften your cheeks, loosen the muscles around your eyes, let mindfulness flood you like a paddy field in monsoon, let the water grow still, until the sky takes up its home there, until all is surface, your mind a mirror that effortlessly reflects what comes before it, without moving an inch towards it; floating on the paddy field, head opening into pure sky, the sound of water lapping gently around you, as though there was nothing to be done, and no one to do it. Relax ... Deeper ... I got you, Jim ..., floating on the surface, let her scream come to you, let it pop out from all the others ... Lay back ... I got you.'

He imagines sweat mushrooming on Jim's forehead, skin turning to grey paste, breath jagged like a hooked fish, as if he'd caught something deep in him. Of course he would have to practise reciting this script, in the one-in-a-million chance their paths ever cross again, punctuating the most significant details with pauses, understatement, different degrees of eye contact, using drama itself to help Jim encode the new memories. It would be nothing though compared to how much Jim would have to practise, over and over again, a thousand times or more until the memories find a way to take root, or something to take root in and slowly turn themselves into new ground.

'Action. Used to being ignored, the body still makes the really big decisions without us. You are already walking. The same you that was driving down Putney Bridge Road, sitting in his car while nothing around it moved, that carefully laid the table, decided to go fishing (the you that we are desperately trying to stitch together), that same you finds itself running towards something that isn't moving, to catch up with it. Children are screaming. Dogs are going nuts. Something is really happening now, you know because, in the moments before seeing has become understanding, the chaos of the scene has magically stilled itself around a focal point, pulled into order by an audience. Kids and their dogs have gathered to watch someone who is strange and unrelated grab someone who is the very opposite of strange and unrelated by her neck. She is drowning and he is saving her life, except there is no water. He is dragging her from a burning building, except there is no fire. He's a toad kissing a princess, except he's not kissing her. He is stopping her from choking. No, he is choking her.

'Jim, we are moving through the door. I should be honest with you at this point: some of these details are mine not yours, because we can only find yours via mine; dangling them in front of what's buried deep inside of you, like bait to draw it out. Are you OK with that, Jim? I need your consent ... Think of it like surgery, where one thing is inserted and something else is removed. Except there's no anaesthetic, because it only works if you really feel the pain's hellish detail. In time you'll forget the details are mine, the stitches will be removed, it will all be of one piece again: you and me, me and you; forever yours, forever mine.

'Running towards the fray you are shouting but nobody hears. He's puny, the boy, but still strong enough to lift her up from behind, holding her so her feet don't touch the ground. This is the moment I warned you about Jim, right now. You are shouting but nobody is listening. Like being in a dream. Like being in someone else's dream. Running towards her, driven by a will that isn't yours, by a body that is barely yours, while the rest of the park recedes like a tidal wave after the wrecking, leaving her marooned in the young boy's arms, a pale white form. She is drowning and he is saving her life. In her lime bikini. You told her not to wear it, that it was only for the beach, like Broadstairs. In your mind she hasn't grown up, still happy to feed starfish with her sister. Some of the audience are watching through their phones. You feel your body turn hard, as though bone and flesh have swapped places. Your fingers have curled around an imaginary iron bar. It is happening. This moment. Full circle. This is reality. Right here. Plant a flag in it.'

He would pause a moment, let silence do its silent work, catch his breath, then begin again in a gentler, less urgent, more documentary voice:

'*The Lovers of Hither Green*. Like a painting. Her face is soft, relaxed-looking, lying on his shoulder, supported by the crook of his arm. Except there's a line of dark blood coming from her nose. Something tells you it's not her, don't trust it. The decision has been made. You are committed now, the rest will follow. Count how many boys there are: three, four, five, six. They are not from round here. They are not from this continent. They belong in an African desert, where it's not as hot as it is here today. You are in control. You have got this. You pick him up by his jacket collar with just one arm. This is what all that training was for. Notice how light he is, ninety pounds, a good catch, feels lighter than you thought. You sense something on the back of your leg tickling, a minnow bite. More of them are appearing, coming through the trees; the more the merrier. You pick up your darling, wiping the blood from her nose, telling her the party is over, you tell her she's safe, you tell her you got this, you tell her that Daddy's got this. You'll retrace your steps home carrying her like you used to, she can fall asleep, snug in your arms. Feel how good it is to be saving her, it's what fathers do. Visualise it ... Then let go of what didn't happen, and feel for what did ...'

He remembers how the physiotherapist would marvel at Jim's routine, semi-professional in its intensity: jump-rope, sit-ups, Russian kettlebells, chin-ups, burpees, shadow-boxing. He'd hit the road running hard around the hospital perimeter for an hour holding guard with his hands. It was the one thing he was intense about, without knowing why. Otherwise he was laid-back, like a chalk outline drawn on the grass, and they were laid-back with him.

'When's the fight?' the physiotherapist would joke. It had already happened.

'Did you feel that Jim? Because something has just been inserted in your head that shouldn't be there. It changes everything. You go quiet, a stopped clock, a young man who turned old waiting for his date under a massive, crushing sky, like walking home on an autumn path towards an empty house. It's confusing, isn't it? Yes, that's an understatement. Remember the moment I told you about? Well, you're on the other side of it now. Don't panic, just listen to what I tell you: I want you to rise up, take to the air, like you're a hawk, hovering high above us, looking down on the mess. You see little figures being sucked in, organising themselves around a focal point – a shiny black blind spot – then exploding out, then running back in again; in and then out, flexing like the lens of a mad eye. On the periphery the little figures are desultory, unmagnetised by what's happening. Your eye is sharp enough to make him out amidst the melee – you, that is – sitting patiently, cross-legged like a Buddha while the demon forces of Mara attack him. Buddha tries to stand up, drunk-looking. It's then you notice something coming out of his head that isn't normally there. Zoom in slowly, at your own speed ... Closer ... Closer ... One of the other cabbies has a cap that looks exactly like it: a foam handle sticking out of the side, technically the temporo-parietal junction, "Has anyone seen my screwdriver?" written on the front. That's what it looks like. "Has anyone seen it?" That's what it looks like he's asking the kids who watch him through their phones and Go-Pros as he staggers around in circles. From this distance, from up here, it's comedy. Everybody has seen his screwdriver apart from him. He sits back on the grass. Then he lies down and stretches himself out, waiting to be drawn around by the police, looking up at the sky to see himself looking back.'

He imagines how hard it would be for Jim to repeat this script, to speak it, over and over again – a hundred times, a thousand times – the same words and images, his face as white as the inside of an apple, limbs twitching like a dreaming dog's; saying them with enough conviction that they find their way home. That's the idea, to give him something real enough – however invented – that the trauma finally has a shape he can recover from.

But then what if by doing this the neuropsychologist was making things worse not better, creating a new tributary for Jim's suffering to rush through? Imagine that: if the story he was telling multiplied Jim's nightmares, gave them new forms to inhabit.

'I know what I'm doing, trust me. You feel cold metal. Somewhere in you there will be a record of what it feels like. Can you find it? Thin rivulets of blood slaloming down the side of your head. You don't believe this is happening. Your disbelief is part of what's happening, as though the character and the audience are in the same mind. The metal is in there too, waiting for you to find it. "There goes Broadstairs." Dry, anticlimactic, by-the-seaside wit. It's the sort of thing you might say. Because even though everything has just been lost in an instant, even though "you" have just been erased by a twelve-year-old boy and his friend's screwdriver, you are still yourself, still a hawk, wavering gently like a hallucination in the streaming air. The football is over, the playground empty. The scene around you continues its cohering, the event's radius growing, now strong enough to organise most of the common around it. You are the event, momentarily the centre of the world, reality's fovea. And you get to watch it too. You see a woman running towards you, Carol followed by a young girl, your other daughter. You see the woman's face change. She turns to the girl and screams at her to come no further, understanding instantly the nature of all trauma: that once the door has been opened it can never be closed. It's too late for her as she falls onto her knees by your poor head; something authentically new has just been brought into her world.

'You see the police arrive and cordon off the area like a cricket square so that everyone knows where the stage starts. There are paramedics crouched over you. Imagine that from up here your raptor's eye is powerful enough to see the catheter inserted into your femoral artery, that your ear can pick up the concern over blood pressure, the debate about whether to leave the "weapon" in situ. Constance, David and Talbot. They tell you their names.' (You will see them two weeks later on the ward when they come to visit. You will not notice the surprise they try to hide when you can't remember who they are. All people see is the handle sticking out of the side of your head with you chatting away. They don't see that you have disappeared along with your memory, that you are no more there than anywhere else now. Still, you say thank you, passing on a bouquet that "Carol" − even she has to wear a badge with her name on pinned to her jumper − has just handed you.)

'More people arrive in high-vis jackets with a stretcher. "Substitution ref ... Someone must be poorly." You can't stop joking. Then you hear the terrible din of splitting air, wind chopped with such sudden force that you are reeling, in free fall, nose-diving back to earth, to where the air ambulance is landing next to you. You hope this one's off the meter. The helicopter doesn't have your knowledge, it relies on landmarks, or satellite images on the dashboard to find its way. You always wanted to see the city from a helicopter, but lying on your back you only see what's above you, the sky turning red as the sun sets. Except your eyes are shut and it's turning red anyway.'

You don't see the audience below un-crane their necks the moment the air ambulance disappears behind the industrial estate, while men in black dismantle the stage, any lingering attention shooed away like spilt popcorn, like news that's rubbish as soon as it's made, until there is nothing left – beyond pressed grass stained red – to imagine something was ever there.

Beyond the park, there isn't the faintest pulse of what just happened. Some opportunist used the distraction to walk into a neighbour's house and leave it moments later with eighty inches of home entertainment. Your girls are back at home by now, sitting in silence at a table carefully laid with cold takeaway. Behind their unlined faces trauma chemicals are making memories, lasting signatures which will be difficult to access, near impossible to regulate, hot splinters of a reality that won't go cold for as long as they live, that like cuckoos will evict everything else.

'Imagine Jim, as you fly towards NICU, that below you, somewhere deep in the warren of low-cost housing, hiding behind a row of recycling bins, there is a twelve-year-old boy with blood on his hands, breathing hard, not sure what has just happened. You imagine this because it will help you. We can do it together, write him into life, the boy they never caught. Imagine him because it's all we have left to connect us with reality. Say, perhaps, he arrived from Yemen three months before, having crossed the Mediterranean in a crowded dinghy without his family, travelled through Europe in the back of meat lorries, held up in dark tents on the French coast. He's seen dozens die in front of him, this boy. He last saw his mother and sisters more than a year ago. If they are alive still he doesn't know it. Imagine what that's like. He has not met you before today. He had never been to the park. His one friend, an older Somalian teenager, gave him the screwdriver in case they met another gang. He's small for his age and the youngest, the one with the most to prove. So small in fact that he had to jump up and slam-dunk the screwdriver in your skull, force it through the hair, bone, dura, arachnoid, cortex, into area CA1-3 of the mesial-temporal area, more direct than a surgeon would dare. He wasn't being precise of course, and yours is larger than average because of the Knowledge, but the effect is precise: a new you with no ability to learn from the future or recall almost anything of what had gone before; pure present tense, a life turned back into just living. And in that instant the two of you are joined: you are lost, and he's lost too; you could be father and son.

'Between us, Jim, we will keep him – this creation of ours – under a glass case, pin it in place, so that we can change the angle of our scrutiny again and again, until the circle of our understanding is complete. It will work. You will understand him because it will help you. And if he should ever come to you without invitation, in the middle of the night, say, like a terrible groaning in your ears, a horror vapour through the cracks in the window, a monstrous small boy, his teeth sharpened to a point, then you shall change too: we'll make you twenty feet tall, with thick reptile skin and a dead voice capable of silencing him instantly. He's ours, Jim. We have designed him, we can just as easily have him turn his weapon on himself. In time, with hard work, we hope you might choose something different for him. Either way, you need him, Jim. Without him, without someone like him, you will remain stuck without a single memory in the moment you were attacked.'

★

In reality Jim had walked in through the sliding doors of A&E on his own, a little pale – no blood, no broken bones – a six-inch screwdriver plunged in his head. There were no witnesses. His first-person perspective – housed in less than a cubic inch of white matter – was lying on the floor somewhere in South-East London. From that moment on he became a story, laminated on cards pinned above his rehab bed, pieced together from reference material: a timeline of his medical treatment (surgery, intensive care, acute neurology, post-acute rehabilitation); photos of his medical team (Mr Webb the surgeon, who took four hours to remove the screwdriver, the paramedics, various nurses, his occupational therapist, the physios); another card with Carol, Hannah and Miley; lists of 'Likes' (fishing, barbecue, Crystal Palace Football Club) and 'Dislikes' (Chelsea, bacon, the European Union); fishing photos, as a child with his father, as a teenager under long hair and an impressive moustache, brushed up for his wedding, on holiday with his kids. Photos to make him feel at home. The crucial one, the one of what happened to him, was missing.

The treating team deducted some things from what the police told them, and assumed others. They ignored the gaps, and the occasional looks of terror that ghosted across his face. At the time it didn't seem to matter because Jim got better anyway. In many ways he was dedicated to getting better; aware of memory loss, superficially adjusted to the new deficit, able to rely on aids (notebooks, phone calendars and reminders), able to satisfy basic goals first (time-keeping, independent dressing, personal hygiene), then more demanding ones (shopping trips in the local supermarket, using a simple map). The motivation to get back what he'd lost was strong. Jim was a model patient: the team had a model and they fitted him to it, which took the place of what was really in front of them.

They left discharge planning late, too late. The idea was for Jim to return home to Carol and the kids. Nobody could see a problem with that. A week before he was due to leave they made a home visit. The neuropsychologist sat in the front of the minicab – the NHS wouldn't pay for a black cab – Jim sat in the back next to the occupational therapist. As they went over Putney Bridge, up to the Wandsworth roundabout and then along the South Circular, he watched Jim through the rear-view mirror, mouthing the names of streets, then giving up. They pass a huge grey industrial estate, a small settlement of travellers' homes where grey horses nose at rubbish, a large sprawling common, a lone wind-blown woman shouting for a dog. He watches Jim scour everything he sees without finding what he's looking for. The minicab follows the satnav through the maze of the estate until they reach Jim's house. There's a parking space outside, and a friendly neighbour, who remembered seeing him the afternoon of the attack. Everything is going to plan.

It's only when walking up the path to the front door that something suddenly hooks him. Bent double, eyes turned all pupil, he's mouthing something but it's not words, he's gasping for air, like he's inhaled something terrible and can't breathe it out. They are asking him to tell them what's happening but it's already happened. He is running down the street, and they don't know if he's trying to catch something or escape from something else, while Carol and the kids watch from the front doorstep. Later in the car, he looks like the ghost he has seen.

They have finally caught up with him; the real patient has arrived at last, replacing the obedient, easy-going impostor of the last eight weeks. The memories haven't gone after all, they've just retreated to a place that can't use words. Now it's too late to do anything about it. Discharge was in motion, the bed manager had no leeway, NHS rules. They have to find him somewhere else to live within a week, it's not safe for anyone if he goes home. Which means a halfway house on the other side of the borough, whose guests were refugee families, wet-brained alcoholics, long-term schizophrenics and young, bewildered killers in the making.

That was the last he ever heard of him.

★

A virus which adds other people's stories to one's own. A symptom of loss and a way of preserving that which needs to be lost.

It was Patrick his therapist who first suggested the journal, that it might foster a 'deeper economy' of thought and reflection. This was years later, long after he'd walked away from his profession for good, and just as he was coming to understand that he could never leave the patients behind. There were endless fragmentary notes about the thousands he'd seen across his career. One note described waking in the middle of the night after a vivid nightmare in which he bumped into Jim years later, but here Jim was the treating clinician, and he the devastated patient. It made him think:

> *Just in case you ever do meet him again, have a story ready, made up from scraps of memory — fishing, car ride home, the daughter's bikini — about how he came to be attacked. It should have enough force to pin him to it, make itself his, so he might finally join himself back up again.*

The more he wrote the more it became apparent that trauma was part of his memories; wherever a thought began, however lightly or brightly, it would find itself caught by its dark undertow. On occasion he turned to face it head on:

> *… It's a perpetual-motion machine with no apparent half-life, intensifying the further it gets from its source. It's measured in feelings of self-reproach, in not feeling anything. In amnesia, but also the failure to forget.*

But then trying to characterise it itself became an ordeal.

> *Its symptoms include restlessness; a need to move on frequently, to avoid it, while searching for places that fit the conflict perfectly.*

They also include making notes about trauma; relieving then entrapping; its own dark narcissism.

He wondered why he had sought out such extremity in his professional life; the unseen forces that brought a doctor and patient together.

Patients aren't selected for their dignity and forbearance. Doctors aren't selected for their empathy or imagination, for their capacity to love ...

He thought of Bella.

Imagine what it would have been to love her instead of killing her. It's still a possibility; one might spend a whole day imagining the last whole day of her life; the moment immediately before what made her a patient happened, in all of its detail.

'Love' and 'Trauma', they had become confused in him; overlapping symptoms, similar-colour maps on fMRI, he imagined. More generally he thought how both words had lost their way, stretched over too many different things, stood in need of their own rehabilitation.

The journal was not just Patrick's idea. In the places he found himself everyone was at it; on Lake Atitlán, in the Sacred Valley, at the ashram in Tiruvannamalai, or by the volcanic baths of the Esalen Institute. Entrepreneurial locals sold pre-aged, leather-bound notebooks, to give those who wrote in them the feel of being august and sage. Patrick was right, it did have a *deeper economy*, it belonged to Jeff Bezos, to an algorithm that could make 'nothing' look attractive: like meditation apps which sold recorded silence, overpriced 'journals-in-waiting' – 200 empty pages – were top sellers on Amazon.

Initially he found the writing therapeutic to the extent that it wasn't reflective. But gradually a tone crept in, a stranger wedged a foot in the door, watching him. Writing turned into a reflection, but of what they – this stranger – wanted; words were written, indeed life was staged, for their benefit. At the heart of his most intimate experiences he found made-up characters, plot twists, facts battered on an anvil into whatever shape 'he' chose. Eventually one notebook was full. He called it a volume of his journal. The publishers called it *Let Me Not Be Mad*, and described it as 'an avant-garde memoir in the form of neurological case studies', and 'a self-portrait in a convex mirror'. Really it was a notebook. Really it was a lap dance for a stranger. Really it was a cry for help.

★

Of the different ways writing *Let Me Not Be Mad* drove him mad, none would be more symptomatic than the abandonment of his own name. He had been the one to suggest it as a way of protecting his patients' identity and keeping his real family from any public association. Patrick said there might be 'a deeper geology' (his therapist always longed for content with extra dimension, just as he preferred casting shadows over content to shedding light); the fantasy of escape, a form of 'self-identity theft'. The publisher seemed unconcerned and the idea disappeared. He was in a chai shop in Himachal Pradesh, the completed text on its way to the printers, when the publisher called him, urgency in his voice. Legal currents can change quickly: that week a judge had ordered a huge payout on a case hinging on the protection of personal privacy (this in a world where having intimate details 'liked' by as many strangers as possible is life-blood, the closest thing to love that people know). He had half an hour to come up with a new name for the title page.

'A. K. Benjamin.'

He could have had anything.

'Really, A*****? Are you sure?'

Anything!

'Er … yeah.'

Try as he might to extemporise a purely fictional pseudonym, he could not break free of reality's orbit; 'Benjamin' was his name for the first two weeks of his life. His mother had a strong intuition over the months she felt him in her belly and imagined his character. She was even more sure of his Benjaminness when she looked at him through the glass of the incubator on the ICU ward where he spent the first days of his life, unheld, untouched, as yet unaware of his vocation in emergency medicine. Classic Benjamin. It was only after she got to pick him up and look into dirty, half-shut eyes – no less shut now nearly five decades later – that she realised her mistake. But 'Benjamin' also because of its Jewishness, the association with giants – Freud, Wolfgang Köhler, Michael Eigen. And because for all of his anti-medical antic he loves the solidity, the straightness, the unimaginativeness of doctors with such stupid sentimental canine force. What could be more unimaginative, more solid-sounding than Benjamin?

Only then, the name gone to the printers, did he check on faltering Himalayan Wi-Fi the real provenance of A. K. Benjamin. Still now, two years after the book has been published, the first A. K. Benjamin on Google, the real A. K. Benjamin, his double, is not a doctor but a shortish, genre-defying, gender-defying electronic cover artist from Taipei; the opposite of original, the opposite of giant, the opposite of solid. With 'alerts' he is notified every time something is posted by his namesake, like when, a few days after the book's release, the other A. K. Benjamin Instagrams himself holding a copy with the caption: 'Decided to write a book!!!?! Long story!?!'

The identity theft was mutual. The neuropsychologist's daughter 'Bronwen' (not her real name either) became the first and only person in the history of the world to 'like' both A. K. Benjamins.

After the book was published, he tried to go back to work at the hospital for a few brief, dark, difficult months. Everybody knew it was him. Someone will someday write a memoir about a doctor who, after writing a more or less honest memoir about his patients, resumes working with them. There are looks; there are looks that get mistaken for looks; clinical minutes are lost wondering what the person on the other side of the desk must be thinking, what might be said safely, how it might sound coming out of a character's mouth, a character with a different name that is still too close to his own, who finds himself in a quasi-fictional supposedly avant-garde case study cum memoir. *He had let them down.*

He gets a call from John – not his real name – a childhood friend he hadn't heard from in years. John O'Neill was big-mouthed and implacable, a shot-caller at school who from his earliest years got to decide what was going on. Maybe it was a consequence of coming from a poor, overcrowded family; so poor in fact that his uniform was always made up of seconds from the Army and Navy store. Usually he could blend in with the motley grey and browns of the school colours but there was a year photo with him on the end of a line of much taller boys, beaming in the uniform of the Japanese Admiralty (his mother had splashed out especially for the photo). Pride without a soupçon of embarrassment or fakery; that's how superior John felt, that's how sure of himself he was. He knew from television and radio that John had become a successful foreign correspondent – moving from one conflict zone to the next – war, atrocity, famine, natural disaster – his preference for combat attire and telling everyone the facts of what's going on seemingly undimmed. And here he was now shouting on the other end of a satellite phone from somewhere, about what a terrible attempt at disguise the book was; he'd sussed him after a few paragraphs. As for the rest of it, John was characteristically direct: 'Not my thing; self-involved, miserabilist, postmodern crap ...' He got the picture. John continued: if only he'd written the 'real' story, documented in straightforward, journalistic idiom the carnage of his personal life, the bona fide private atrocities, the domestic natural disasters, *the facts* that is, the book would have been so much more interesting, so much more compelling, so much more original; he got that picture too. *He had let himself down.*

★

It was time to pitch the sequel to the first book. He wanted to '*renegotiate* ideas about trauma', he wanted to '*resuscitate* the word love' and, following John's injunction, he wanted to make 'the real his quarry'.

Weren't they, after all, the same thing? . . . In a way? It helped if he wrote such questions with a mystical whisper in his voice . . .

The book would be a case for love, half a million carefully wrought words, beginning with a doctor and his patients but then stretching over various unlikely occult terrains . . . *Because every relationship involves love somewhere, sometimes, if only momentarily. Perhaps the word itself is a branch of all relating?* (There might be a Venn diagram for this bit.) *For many it's all they have in the face of meaninglessness and suffering, and that's almost nothing at all: the only thing that is real, while at the same time being an alias for what we can't handle about reality . . . A password for what we have left to pay attention with 'in these times'* ('these times' always rang a bell, whatever 'times' it was) *and for the quality of that attention.*

It nearly made sense. The publishers *quite* liked it.

Originally the book began with the straight telling of an old story, a true story, the kind of story John wanted him to tell. (Also the kind of story he was worried nobody would believe, the kind of story he was ashamed of.) It was summer 2019 in Varanasi when one afternoon he sat down at his guesthouse desk and cast his mind back to England, mid-1990s. 'A. K. Benjamin' was a failing screenwriter in his early twenties, addiction issues, girlfriend problems, father complex; no kids yet but a brain-damaged dog that depended on him. He got a call from his expat brother about a childhood friend who had suffered a sudden catastrophic illness, a virus that had attacked her respiratory system leaving her ventilated on an ICU in Bristol. Would he mind taking some flowers in on his behalf? It was the first time since his birth that he'd been anywhere near an ICU. Strange, given how much of his working life he subsequently spent there. Stranger yet given how central respiratory issues and ICUs would become a few months from the time he was writing, like the future calling.

Once upon a time in Intensive Care:

From those first moments he was entranced: the silence, the heavy doors, the smell of sterilised floors, the minimal music of the ventilators, the lost urgent look of families waiting. Leaving flowers – limp daffodils not yet in flower – with a nurse, one of her close friends overhears and introduces herself. He explains he knows her from their school days, but hadn't seen her in years. A thrill to hear himself fabricate so effortlessly, without intending it. The friend trusts him; why would anyone lie. He asks about her 'prognosis'; 'it's critical right now, too early to say anything, just "watching and waiting"'.

The next day he buys a clean white shirt, polishes his shoes and gives himself an accidentally severe haircut in front of the mirror. At the ICU door he hands another nurse some flowers, two dozen roses this time from the upscale florist opposite.

'Roses – in Intensive Care – get you!'

'How was her night?' Sombrely, learning to speak the right words with the right tone. The nurse unconsciously reciprocating with her technical update: oxygen saturation still at 60 per cent, haeomodynamically stable.

It was Day 28. Even if she made it, there would be devastating consequences. He sits briefly at the foot of her bed, watching her unwatched: ventilated, air becoming breath, feeling every inch of her in his mind, imagining as the afternoon slowly darkened what it was that had brought her here, until a nurse gently asks him to leave. Back at home he looked up her condition: 'begins as lung infection, which can become pneumonic, causing respiratory failure and coronal compromise ... Vulnerabilities apart from smoking and diabetes, include acute stress and emotional trauma ...' She has had her heart broken. He knew that, could sense it from the foot of her bed. He calls the friend of hers he'd met on the first day to update her: '... more of the same I'm afraid: the next few days are critical ...' Pitch-perfect, born to it.

By the time she left ICU – her heart and lung self-supporting but fragile – he had become a central hub in the daily round-robin calls; not quite high enough up the ladder to speak directly to family, but on the next tier of close friends, assigned other less close friends and colleagues to update. Because of the disparate nature of her group, the lack of social media, the general credulousness that the situation lent itself to, plus his talent for impersonation, his doctorly instincts; nobody questioned him. (Years later Patrick would explain that it wasn't being a clinician so much as crisis and subterfuge that were the neuropsychologist's true vocations, finding 'a deeper dramaturgy' in a theatre outside of himself, one that mapped his inner drama.)

Life, the true artist, was writing the screenplay he had failed to:

She returns to her family home for rehabilitation under a community team. Several months pass. All that real feeling that passed between them while she lay unconscious, had he just imagined it? Then one morning he receives a text from her introducing herself and thanking 'P's wonderful brother' for all the support. She hopes to say thank you in person some day. He writes back, heart pounding, 'a privilege, your fortitude, dignity, beauty of spirit have taught me so much about humanity, the fleeting passage of our allotted time ...' sounding like a nineteenth-century Romantic novel, or more precisely like the milquetoast Linton from that novel which he won't read for more than another decade. He'll be visiting his

parents nearby in the coming weeks. If she'd like he could take
her for a drive.

The satnav takes them to Haworth, home of those dog-obsessed
sisters, who knew that romance was always hemmed with savagery.
She's incredibly frail-looking, tubercular, wrapped in a huge bear
coat in late spring, stepping out for the first time from a long,
sleepless hibernation. Ramona, his dog, is excited and jealous,
nearly knocking her over as she tries and fails to stand on useless
rubbery hind legs. Gentle, decorous – even as her cheeks, white
as an envelope, are licked – she's everything he might expect her
to be. He notices the small scar where her tracheostomy had been;
some things will never heal.

Love and Trauma. It happened a long time ago and it was
still happening.

Is that real enough for you John? … Is it? … My diminutive reality instructor! …

Not a word of it made up. One minor lie had cast him
as the lead in a Gothic psychodrama which could only head
towards darkness.

Writing it out was a kind of waking up, only now he finds
himself inside another dream of sorts; supposedly writing this
first chapter, but instead watching a 24/7 live-cam of a sea-eagle
mother in Oregon going to the ends of the earth – at the very
brink of starvation – to keep her chicks alive. (He didn't know
there were so many unshed tears in him.) It's the middle of the
night, the chicks unable to sleep without their mother who has
been away fishing for the last four and a half hours (he's so
gripped he's forgotten to eat his vegetable thali), when an email
arrives from his publisher asking him to review a neurologically
themed memoir for the *Times Literary Supplement* in which the
author reconstructs her emergence from locked-in syndrome
following a brain virus. He writes back saying he's too busy,
but the publisher immediately reminds him he's contractually
obliged to get the 'A. K. Benjamin' brand out there.

He minimises the live-cam for a moment to read the blurb: 'The extremity of the condition and its consequences render the narrative so compelling ...' This was code: it meant that the author, like medical writing in general, treated readers like trauma-starved children, forcing her way down their throat with her long beak so that everything that came out of her mouth – regurgitated, shocking, indigestible – goes straight down yours. 'My reality,' she might say, 'it's more heightened, more concentrated than yours will ever be, anchovy to your fish fingers.' As he scrolled through her story it wasn't the horrible brain virus or the lassitude and ignorance of medical opinion that bothered him. To be honest she could have played those more heavily, eked out the drama, and it would have been OK by him. What made him balk was the author's relentless self-positioning as the sworn enemy of doubt, loss, introspection. There were moments he felt sorry for the illness; how could it hope to thrive in such company? On the live-cam it was dawn on the Oregon coast. There on the edge of the nest set in the crag of a giant rock face, two eagle chicks – so young the down is still fluffed around their necks like choristers' ruffs – were standing, shivering with hunger and cold. Their mother, who spent the last weeks getting them to this point, had died in the night. They were alone under a massive, crushing sky, as the sun edged into frame. He remembered that holiday in Wales when lost on a crowded beach, the tide rushing in, he watched his father – the man who eclipsed his childhood, of whom he is still not free, years after his death – running round like a maniac trying to find him, and how he sat there watching with an ice cream. Love is trauma isn't it John? Really we are all orphans on the edge of the world. They stared directly out of the frame, looking for their mother, looking towards me. I thought it was fear and grief that were making them shiver. A moment later as the sun cast them in its warmer light they looked like young lovers, trembling with possibility, wings touching as though holding hands. If they didn't try to find food it would be their last hour on earth. It looked like they had jumped off the rock face as they fell like feathery stones, until the moment falling turns to flying, swooping upwards, side by side, over the beach and ocean below.

It was only then, the young birds soaring over tall trees and canyons, that the former neuropsychologist finally woke up to what was going on, recognising that what Patrick would call a much 'deeper neurology' had been leading him via its cryptic map (the girlfriend in a coma, the locked-in memoir, the orphaned birds turning Chagallian lovers) to Bella, the woman he would end up calling 'Bella' who the nurses wanted to kill. She had died around the same time as his father, and it made those few months, both of them so ill, feverishly unreal. He couldn't go to her funeral because it was on the same day as his father's, but he heard about the celebration, the hundreds that showed up from different chapters of her life who hadn't been allowed to visit; the colour, the music, the food, the general exuberance, that could never have been guessed from the last grey months of her living. Then he forgot all about her, reburied her in effect. Only in trying to tell the real story of his life – prompted by Rear Admiral John – he'd been writing about her all along. From that point on he allowed her story to overtake his, and make imagining her last horrendous weeks the first chapter of the new book.

But it was still not quite real enough, was it John? Because in making this decision, in writing that first chapter, much as he was opening to someone beyond himself, he was also shutting down; editing out the anger that all he could do was watch her, that all he could do was write about that kind of helplessness. Only now with the chicks facing the next life-threatening task of teaching themselves to fish did he see that there was something more real, more honest that was still required of him, a confession that had been beyond him until this moment. By writing himself up as another character, camouflaged as a 'third person', he ensured that he could never come out from behind the observation glass, could never fully inhabit himself, could never really touch her, never turn from profile and look beyond the frame, straight into those beautiful limpid violet eyes and say:

'Bella, I love you.'

A beat.

'And you are not the only one.'

★

The POV is at the level of the handle, which turns, and the door opens onto a huge American kitchen framed between yoga-panted legs standing in front of us, our nose at crotch level. (A juvenile fantasy of male programmers no doubt, neither relevant nor appropriate in a clinical tool.) Of course you are not interested in that, preferring to track a late summer bumblebee which sputters past, a foot above our head. We snap at it but nowhere near as quickly as either of us would like, the thought preceding the actual snap by a lifetime. (I imagined quicker reflexes for us.) Up close the bumblebee looks fantastic. The legs have bristles that themselves have bristles; the front pair specialised for cleaning, the rear for pollen collection. There are even those thin dark veins, with capillaries branching from them, patterning the wings. It's pellucid-looking – biologically accurate as far as I know – in keeping with your extra acuity. It's also a pet love of the design team, I imagine, as I note how far more care has gone into rendering the bee – the kind of challenge coveted by the more creative members – than, say, the equally challenging Anaglypta wallpaper, or the scuffed, goosebumped knees of the children under the table, one of which the bee has just settled on.

'Get outta here, Benji, you're filthy,' says an adult American female.

The direction of the voice isn't quite right. It's not terrible, it could just be better. Benji must be our name because our body is leading us towards an exit. Except there's a hint of resistance in leaving; we're hungry and the yoga pants are serving chana puri for breakfast.

'Bhu-bhu-bho-bho,' we say. This must be the Indian English language version – our barking has a Marathi accent.

We go through another door which looks like it will lead to a lounge or a dining room, but when we go through it (our tail lashing the frame with a high-quality wooden noise) we are surprised by the white fission of sunlight dissolving credibly onto the scene of a beautiful spring meadow, a beautiful art-school meadow, like something one might see in a film by Satyajit Ray, witness the fitful flight of butterflies, the whorling swallows, and – nitpicking – their slightly stacattoed dives. We are moving through grass which parts convincingly before us as it flattens in our wake. We know what we are because of the cues and context around us: the end of our wet tongue flapping in the bottom of the frame, the tip of our nose fused with the horizon. But really we are nowhere, pure first person, the space in which everything keeps happening. Not even the contrail of a self, because the self is only ever contrail.

The POV moves like the snout of a gun dog trained to the ground, zigzagging over the earth for its quarry's scent, then stops suddenly, as we pick our head up, holding very still, pointing. A pointer, that most philosophical of dogs, trapped mid-distance between subject and object. At the level of experience smell is only inferred from sight. We should be drowning in different odours in a meadow like this, but all I smell is the moulded plastic of the headset. Then, out of nowhere, in the middle of the field stands a barn and a little too quickly we are facing its door. How do we open it? We expect to go through, I feel how dumb we are, how difficult transitions can be. It's such a winning touch when without thinking we raise up on our hind legs and paw the handle, nosing our way through the crack.

Once again the new scene is not what we expect – the inside of a barn – but a corporate boardroom in Hong Kong, mid-session, a CEO barking orders. The conversation halts and the suits look round to register our presence in the room. I am as shocked as they are, and it affects you as we lower our head, our breathing quietens. Somehow this body, this POV, is giving me the feeling of sheepishness – its doggish equivalent anyway. I want to apologise, I want to introduce myself, I want to find out about what's eating the boss.

'Bau-bau,' Cantonese for 'woof', is all that emerges from the prison-house of language.

Is this what a dog feels but can't say? Or does the bark give it the feeling that it has spoken sensibly? Has any of this – the cross-species interface, the exact ratios of dog and human, this split-brain werewolfism – exercised the programmers enough? I feel sad, the knowledge that even with the most scintillating technologies we will never really know what it's like to be a dog. Of course it makes me think of Davis – just about the first patient I ever met in my professional life, then of patients in general who would need to be understood, of abandoned pets, of lovers whose eyes went unmet, of babies who couldn't be named.

We approach another door that leads out of the boardroom when, for no reason, there is a jump cut from first-person POV to a third-person dog face: our face – if we stitch the two together – which is really your face.

Except it's not your face – at least, not the face I imagined for you; it's undignified, stupid, even for a dog. I blame you – to myself, that is – which isn't fair.

Then we are back with the POV, moving through the door and another dramatic jump (what is it about transitions that so confounds programmers and lovers alike?). To the inside of an abattoir: giant carcasses hung on hooks, men in bloodstained white coats, the sound of angle grinders, the empty innuendo of overpowering smells, the sight of our breath hitting the cold air, the faster thud of dog-heart. We are given the implication of fear, put in us by them and made our own. Picking up speed, we take the nearest exit – an electronic weight-sensitive door – opening suddenly onto the antiquated living room of an elderly couple on a sofa in front of a wildlife programme on television, craning their heads a little when we pass before them and obscure the screen, mingling momentarily with its footage of a huge pack of dogs surrounding a frightened-looking man. Clichés, I think. There is not enough care, not enough imagination; like doctors, running out of ideas; like doctors surrounded by patients, running from them in fear. The thought that we have to convert what's real into a desktop of reality, a patient into an icon, otherwise it will overwhelm us. More speed, the sound of blood pulsing, pushing through another door, back into the American kitchen.

'Benji! Get outta here!' The same line on a loop, back to where we started, like we just hit the edge of the dog's known universe.

'Benji! Get outta here!'

I wonder if our paws have brought in abattoir blood.

'Dr Benjamin?' The gentle adult male voice with an Indian accent has no possible co-ordinates in the scene. I am determined to get at that puri.

'Benji, you're filthy!'

We pass through her legs, which pixellate terribly, and through another door, and towards another face, younger, open, smiling ...

'Dr Benjamin, if I could—'

The sensation of speeding up, of wanting to stop but not being able to ... Another door, another woman, waiting for us, as we go past her and through another door with its woman, and another, and another—

'Dr Benjamin? If I might tear you away from that headset for a moment? I think you have the wrong software running.' But we are all wearing headsets, always, even you John. The software is not wrong, but to the point: a metaphor for hopelessly divided urges; an increasingly frantic, thoughtless, bestial, search; opening doors indiscriminately as you look for something beyond your understanding: the One, or the One Code; the algorithm that stitches together all the things we love and have loved.

'I think you might be running the dog simulation.'

'Bhu-bhu-bho-bho.'

Like Bella's cry for help turned into an unanswered machine. Like Davis trying to say the unsayable to his wife. Like Jim surveying the strange, unfamiliar city while he mouths the street names, walking up the pathway to his home until the moment he realises it has turned into somebody else's house.

'You asked about our stroke rehab software?'

Bella, Davis, Jim. No body, no words, no memory. The three monkeys. A history of love as a history of what I failed to love.

'Dr Benjamin, I'm going to switch you off.'

I imagine my ears pricked hard, my nose dry, fear in all four eyes.

★

'Some traumatised people invade a school and slaughter children, others will become healthcare providers.'

That's from a book I'm flicking through called *The Challenge of Being Human* by Michael Eigen left by someone – someone who perhaps couldn't face that challenge, or had met it and moved on to another species – here in Khushi Paying Guest House, Varanasi, a few miles from a stroke hospital where I now volunteer from time to time and one of the first medical centres in India to trial new virtual-reality rehab software. On the jacket of the book, an aged man in an ochre dhoti with irradiated white hair stands on the bottom of the steps leading down to the filthy Ganges, the same river I see when I look out of the window, where fishermen are now pushing out a fragile wooden boat past water buffalo half sunk in the mud. Through brilliant distant eyes the man is looking into a whirl-pool a few feet from the steps. Of course he reminds me of the man renamed Davis, which means Davis is likely to look a little more like him in the next draft. He also reminds me of Bella on the edge of the swimming pool, the moment before her brain haemorrhage. The fishermen find their way into Jim's story too, even the water buffalo are included in the bit I'm working on now. A banner pops up on my phone, which is positioned next to the open book: 'In meditation we are invited to still the whirlpool of our lives. We quiet the mind, releasing conjured stories and fantasies. When the waters are still for long enough, we see our reflection.' It's my 'Daily Dharma' courtesy of a Zen Buddhist priest from California called Zenju Earthlyn Manuel. I have an idea: that my trauma script for Jim, like the appeal to Bella, like the initial framing of Davis should carry the quality of meditation. And just as I decide that and begin the work of forgetting it another banner pops up, this time telling me 'Fake Love', A. K. Benjamin's 'new cover' song, will be released today and I think *He's beaten me to it again, the talented bastard.*

This is a factual account of what has happened over the last sixty seconds. The world is thrumming with strange attractions, enigmatic assignments, latent codes. Actually, it's none of that. It's just reality unfolding, while folding on itself, endlessly; going out and pulling in at the same time, like a whirlpool; new events are born – perfect, untouched, original – only to be understood in terms of what already has meaning.

It's now ten minutes later. I meant to use an app to meditate on the whirlpool, but YouTube's algorithm seduced me into watching footage of stupendously large cranes falling in slow motion across cities of the world, and as they fell I saw metaphors for cerebrovascular events, for doctors keeling over into their patients' beds, but also somehow for the removal of unnecessary support, for the end of all construction, for truth. Technically these were metaphoric fantasies but, more technically and more horrifyingly, the algorithm was sculpting my brain as I watched, creating my future self while I looked on dumbly. This is not a metaphor; every single moment of interface my wetware is being hacked, altered, re-weighted; *sculpted*, in other words. Now is that literal enough for you John O'Neill, war correspondent and hopeless addict of military attire? Am I in the right ballpark of the real, my short, outspoken guide? ... Because it feels like it to me. Because, to me it feels too real, like I'm free-climbing on the rock face of consciousness, my feet cupping the thinnest of ledges created by the meeting of two geologic plates – conditioning and expectation – only when I look down there is no fucking ledge John and what's keeping me up here is only a trick of perspective, my face pressed against a cold hard surface which is really just the floor, crushed from above by your paramilitary jackboot. I wish it felt less real John, I really do. If only I could just relax a little, let go and fall back.

Every moment an opportunity to get in touch with reality is squandered. And yet there is no reason that the next moment can't be the moment when the wheels of *The Case For Love* – a journal that became a memoir that turned into a screenplay which was interrupted by a book review and an ornithological live-cam until it was hijacked by the memory of a real patient on life support which was inadvertently railroaded by a memoir again that constantly risks falling to pieces like a giant slow-motion crane – finally kiss the runway, and a new strain of the real begins. It's the moment as the sun comes up that I, a sleep-deprived Scheherazade, fall silent, finally exhausted of stories to tell the rapacious caliph, and, clearing my throat, begin at last to tell the story of Scheherazade, begin, that is, to make my case for love ...

★

Adora Desconsolada and I are planning our trip to the far north-west. Adora is a performance artist from Colombia, and a techno DJ, and an analogue photographer slash Ayurvedic cook. She really is. She describes her work as similar to that of a fungus, in that it 'challenges existing classifications' providing 'intoxicating, alien-like, occasionally fatal alternatives to the dull orthodoxies of meaning'. She really does. This is her first exclusively heterosexual relationship in a decade. In the months we've been together I have come to see myself from her perspective: gentler, more feminine. She has an idea for a new performance. There is a place deep in the Nubra Valley of Ladakh, on the border of India and Pakistan. Nearby the famous Line of Control runs between the two countries – with Chinese territory in sight a few miles away. There is a town there, Turtuk, which is a Muslim outpost in a culture which is otherwise Gaudi Hindu and Tibetan Buddhist. In the middle of this desert, walled off from the treeless, unrelenting mountain scree, the town is a miracle, a small, verdant oasis of fruit and vegetable terraces, pastel-coloured houses, a solitary minaret, men driving beautiful wild-eyed long-haired goats, women eating dried apricots in psychedelically embroidered sarees. A utopia, a magical-realist's fantasy – Adora is Colombian to her core – and, she tells me, a locus for pure, androgynous imagination, unfettered by gender concepts; a place where mushrooms can do their work. She wants us to scramble up the steep escarpment of the mountain above the village, armed with a picnic of feta cheese, dried fruit and flatbread, enough to give us the necessary energy for our non-stop living installation. Once we have crested the ledge above Turtuk we will be able to see the soldiers from the three different countries, arranged at their different posts, watching one another through telephoto lenses. Their job is to observe the other, day and night, in case one of them attempts to steal back a little land and thereby enlarge their nation states. Pure masculine engorgement. Territorialism, in other words. We will use the angle of their sight lines to establish a midpoint which will stand in for the invisible Line of Control. There, on the boundary of these phallocratic warring entities, above a decontextualised feminine fantasy village, at the threshold of fact and fiction, of my autobiography and Adora's, right on the perineum of antiquity and modernity, of East and West, she will ask me to drop my unisex dhoti, get down on all fours, doggy-style, and fastening on her favourite chrome strap-on – which resembles nothing so much as a long-range warhead, stencilled by her with images from different myths of torture, execution and fruition – in the knowledge she is being observed by triangulating Nuclear Forces, she will perform an act for/against/in/on a territory which really belongs to nobody, an act of colonialism whose location is so precise – aggression and comedy perfectly harmonised, my own private Pyrenees – that it might diffuse the potentially world-ending geopolitical stand-off.

'Just relax ... deeper ... really let go ... deeper ... a little more.'

I realise how all this sounds John, like it errs on the side of abstraction, is too cartoonish to be credible, but this is exactly what Adora said, and this is exactly what she envisaged, describing it so movingly from the calf/cow asana on her yoga mat in the Khushi Paying Guest House, in broken English, pausing a moment to sip Ayurvedic tea and hope aloud that however hard the soldiers look, through whatever technology, they will not be able to see because they have no concept for what is happening. In this way they will be like those other native Colombians who could not see the Spanish boats arriving because, like mine, their organs could not receive. She has it all planned out, as she describes it, returning once again to her mix of charades and yoga. On the desk by where she sits is a small spiky cactus plant she carries everywhere, and a textbook open at the Jyotish astrological system, next to which incense floats upwards wrapping itself around her words, her signing fingers. It happened like this John, my oldest school friend; honest it did, cross my heart, not a word out of place. I can still see her doing it.

Yet less than three months later, three months during which I have not written a single word, I am on the other side of another door, lying on Esme's bed in an adobe casita in New Mexico, crying openly. Her daughter Niyeli, who is two and a half, is due home from nursery any moment. Niyeli loves me because I am the only person she has ever met who knows how a microwave talks. It was surprising to both of us, deeper, more plaintive than you might expect and obviously very distinct from the carrot grater, while lacking the vernacular range of her sock collection. As far as she's concerned I know what it's like to be the whole inanimate universe. Next door Esme is stifling sobs, hurriedly bringing her phone call to an end.

An hour ago, hearts set on eternity, we were floating across the infinity pool in her garden (landscaped expensively with native beavertail, yucca and century plants to look like high desert, fenced off from the real high desert beyond), soaking after a long walk with Kojak, her Mexican hairless, over the mesas. That moment somehow embodied our future. That moment was the end of the world I'd been hoping for. Her long white limbs starfished on the surface, the water lapping noisily under her armpits, her chin, the base of her breasts. She ate raw when she could. She mainly stayed out of the sun. She took supplements; macca, spirulina, chlorella and something called Alpha Brain; the four immortals, she called them …

John I know what you're thinking: that the function of these lovers as structural elements eats a little too far into their flesh-and-blood reality. But I'm just the messenger, this is her flesh, this is how Esme displays herself. She is the real thing and I have never been in love like this before, the love that makes sense of, no, the love that redeems all the others.

It shouldn't have been there, that object in her abdomen that I saw the very next moment, jutting out between her hip bones, the seed of a new future that half an hour later she would compare for the oncologist on the other end of the line to the size of an avocado pit. Half an hour after that, now that is, she's finishing her phone call to an insurance broker, trying not to slip through the cracks of a policy that had hedged itself on the good feeling she had always had about her '*bo-dy*' – said like it really did belong to her, only not for long. Mortality in the infinity pool. My love has made a patient of her. I have finally found her and she is disappearing.

On it will go, face after face, scenario after scenario, different doors, different types, each more or less ultimate than the last, down a corridor that gets narrower and will never end, driven by the fantasy of being able to embrace the reality of another person, driven without knowing it by the mirage-thought of a last door, behind which there is finality, bringing coherence to everything that has come before, the secret pattern that threads the eyes of everyone you've ever loved, distilled in a single entity that has been waiting patiently on the other side, until the moment the door opens, ever so slowly, ever so theatrically, to reveal not death but ...

A. K. Benjamin, the real one. As in everything else he has got there before me. I am lying in an unshared hotel bed listening to his 'Fake Love' cover, his prophetic new opus, catching myself doing exactly what he – in his soaringly high electrified voice – describes himself doing, what the original artist was doing: scrolling through all the names on our phones, deep in the middle of the Taipei/New Mexican night, lost, alone, *hella faded in my zone, sick of fake love fake love fake love.*

★

It's just turned 2020, the Year of Failing to Embrace Reality, in California, the Place of Failing to Embrace Reality. I liked Amara on the app and, unusually, she looked exactly the same when we met at a Coffee Co-op on Shattuck in downtown Berkeley. Hair silver and neatly bobbed, more like mercury than hair, bronze shoulders shown off in a sleeveless t-shirt, round and hard as apples. She noticed me looking at them; her 'Patrick' told her to do a handstand whenever she felt upset, she explained. 'A lot of upside-down time over the years. Whatever falls out of my pockets is stuff I no longer need, was never really mine.' It wasn't the near-nonsense of what she said, but how she leaned into it, exposed herself, and in her fragility still kept coming forward to meet the other person, me that is. We talked up a storm, the chemistry was obvious. Then she told me she was micro-dosing on Psilocybin − it was that kind of chemistry. A familiar key change; dark, difficult waters, forty-five minutes into our relationship, her whole look and feel transforming right before my eyes, talking now about the same things from a distance, as though laminated like a self-help manual, making her 'mental health' sound like an accoutrement, or the name of her postmodern pet dog:

'Stop that,' she says.

'What?'

'Turning me into something.'

She looks so sad.

'I'm not ...'

'I don't look sad. Your lips move when you read me.'

'They ... I don't read like that.'

'And stop it with that doctor "look". It doesn't work on me. I've had more electric shock therapy than you've had reheated dinners.'

'I ...'

'I haven't.'

'... I was listening.'

'To the sound of your own voice; oh so knowing, frightened.'

'...'

'A moment ago it was telling you I was one thing, the next I'm the opposite.'

'I've had a really nice time but I should—'

'*How foolish his aim had been!*' she was declaiming

'?'

'*He had tried to build a break-water of order and elegance against the sordid tide of life without him and to dam up, by rules of conduct and active interest and new filial relations, the powerful recurrence of the tides within him.*'

'?'

'*Useless. From without as from within the waters had flowed over his barriers.*'

Silence. There were no anecdotes left. Nothing happens. She was there, right in front of me, unavoidable. A surge of fear, but the nothing keeps happening; her shoulders, her hair, they keep happening. *The waters had flowed over his barriers.* Tears. My tears. When I looked up she had taken to her hands, feet dangerously close to the ceiling fan in the middle of the coffee shop, none of the acutely comfortable Berkeley types giving it any notice. Then the James Joyce fell from her jacket pocket.

'Guess it belongs to you now,' is what she said.

In the weeks that followed we became close and after considering it from all the angles, I agreed to join her on a mission of self-annihilation in the hills.

As we drove we talked intently, the same force applied to whatever the subject, which would have been confusing except we were tacitly collaborating to avoid thinking about what was on its way. The night before I had spoken to Luis on the phone at his insistence, a screening interview. His voice was soft, whistling, ancient, exotic like the meso-American desert where the toxin we were about to ingest is harvested from giant toads. It was reassuring when like a doctor he used Latin names, spoke of 'molecules', dosages in milligrams, tolerances, sensitivities, safety protocols, all in a light, winsome Hispanic voice. It was also like a doctor in that he continued to speak for five, ten, fifteen minutes in the same monotonous tone, letting me know how much he knew about the neuroscience and pharmacology; an interview – typical of certain doctors – where the interviewee is not asked anything. Supposed to alleviate any concerns, it made me more anxious.

Amara's truck pulled up outside a heavy electronic gate built into a high wall made of adobe. There was a yellow sign showing the head of a pointy-eared attack dog. The gate opened noiselessly on to land with four or five highly archi-tected wooden homes nested among tall cypress and eucalyptus trees. Teslas, Subaru Foresters and a golf cart were parked on the forecourt: if these were hippies they had made sound investments. Walking round the compound there was a mobile sauna, two pools, a hot tub, wind chimes, sound bowls, trays of pyramided lemons and avocados, vegan jerky, kombucha on ice. Luis emerged out of this, glowing in a white prana onesie; a surgeon from the future, looking thirty years younger than he sounded on the phone and sounding un-Hispanic; 'Lewis not Luis, from Utah.' After taking a sauna together in silence – the window a stained-glass, rose-wreathed Grateful Dead skull – Amara left me to join Lewis in his Temple, a wooden shack on the edge of the property.

I walked around naked for a while. I drank a kombucha. Then another. I bit into an unripe avocado. I bit into a lemon to take the taste away. I was bored. I thought how boredom had become generally outmoded, swallowed by the urgency, the 'time poverty' that drives the world. Well, boredom was making a comeback, for me at least. Unless it was fear masquer-ading. They must have been in there an hour already, I guessed, while I jammed the digital controls on the hot tub by maxing the jets simultaneously, my body starfished to hit every pleasure centre. The machine sounded like it was going to blow. A long-haired man in a purple sarong came out of nowhere followed by a small, shaggy, indifferent-looking dog, the spit of his owner and nothing like the pointy-eared terror on the yellow sign. The man spent a few minutes pushing buttons, trying to calm down the tub. 'You've cast some spell on it, dude.' He pulled the plug. The tub drained slowly and I grew cold.

Amara emerged from the Temple door blinking, teary, bewildered, like she had just come through the arrivals gate of an airport, the doors of NICU, of life that is, still uncertain which side she was on.

'I have two things to tell you,' she said. 'First, I love you.' I could see she was trying to mean something new. 'Second: don't think of writing about this, you'll ruin it.' I made a mental note.

It didn't matter *what* she said: what mattered was that Amara was still on this side of life, walking, talking her signature near-sense, not writhing on the floor leaking cerebraspinal fluid from her nose. Lewis beckoned me towards the Temple door.

'Oh and three, you're gonna lose your kids . . .' said Lewis winking at Amara, the comedy equivalent of a surgeon saying it shouldn't hurt too much, some of the anaesthetics are quite effective, as he wheels you into theatre. I could just about read some poorly spelt handwritten instructions for CPR stuck to the door as I went through.

The Temple was a small room, the size of a chapel hospital, with an apical roof where networks of dream-catchers hung either side of a blue fluorescent strip. At the far end was an altar on which dozens of toad figures of various sizes carved out of wood and stone and iron were waiting. What looked haphazard at first glance took on a complex choreography the longer you looked at it; like some atavistic CCTV, the gaze of the toads' jewelled eyes covered every inch of the sacred space.

Behind the altar was a large wall-hanging; a face split on its vertical axis, one half human the other toad, smoke pluming from the corners of its mouth. Immediately surrounding the face were other faces, human, then animal, and around those faces the scale suddenly turned universal; cityscapes, skies, planets, stars, the milky way, all of it was innervated with plant life, like a giant garden saturated in typical psychedelic colours. *Psychology, sociology, cosmology, botany; really it's all the same thing, baby* – is how I took it to mean.

'I want to run you through the mechanics of the ceremony, take the thinking out of it.'

Like a surgeon again in the becalming attention to detail. Like a surgeon too in his leveraging of medical vocabulary: 'You might *purge* in this bowl, *expectorate* in this one – in the unlikely event they're required. And there are trees for *micturation* out back.' But most like a surgeon in that there is no real hint of what is to come, of everything that will happen when the lights are out and you're not there and he has the untrammelled freedom to tear into your dura with unbrushed teeth, sweat dripping from his brow into your gaping head, roaming through its greying corridors with fumbling, hungover hands, what-the-hell might as well carve his initials on the inside of your skull before he brings you back. I have seen these things. But I have not seen a surgeon pass his pre-surgical patient a home-made cigar as he chants dark, ancient syllables, waft dense acrid smoke over his own hair and mine as though it were water, and then offer me a regular crack pipe, loaded with a five-mil dose of toad venom, itself loaded with 5-MeO-DMT, one of the world's most powerful psychedelic molecules, a mini blowtorch primed below it:

'Are you ready, Doctor?'

I haven't taken drugs since I got clean twenty years ago.

'No.'

This is really happening. The toads on the altar loom in my field of view like anxiety, shame, escape – old as a toad.

Don't fuck with us, is what they seemed to be saying in their silent anuran tongue. I pause the pipe, make Lewis Patrick for a moment, then my NA home group; it takes me five, six, seven minutes to really layer and nuance my 'share' about the causes of my drug and alcohol abuse, Lewis nodding in the wrong places, looking at his expensive watch, another clinic starting on the hour I imagine. I know what it's like when patients feel the need to detail every last movement of their mind, the fantasy of being completely understood, that there is something solid to understand, something solid that does the understanding.

Still they are looking at me; there are real toads in imaginary gardens.

'I know it sounds silly but if you don't mind turning the toads to face the other way.'

It takes him minutes to reorient the toads, including covering the one on the purge bowl.

'OK, are you ready?'

'No.'

I am not ready. I have the thought that what's about to happen is already happening and that I should share this thought with him, but when I start to report it I find myself already in the middle of reporting it and therefore that I am just repeating what I have already said, and I have no idea how the rules of English grammar are allowing me to draw on them in meaningful speech at all while at the same time he is nodding in agreement with some nonsense I am about to tell him. Which means I have already inhaled the five-mil dose of amphibian poison, and now the room is spinning violently, my viscera have collapsed, and my heart is going into arrest. No problem. I need to get the instructions from the door, which is great because I have them in my hand, right in front of me, only thing is reading them when they have been written in upside-down mirror-writing by one of the two surgeons in front of me who is looking confused, the one that looks like me.

'Are you feeling like you might purge?'

Someone is shaking my head. On the phone last night he told me to have a light breakfast, a 'dieta' he called it: white rice, curd, hot water. I projectile vomit all over the Temple floor: eggs, hash browns, bacon, chocolate mousse, blue cheese, on a giant tidal wave of kombucha, unripe avocados and lemon peel. My 'diet'.

'Sorry.'

I think I hear the surgeon tut. I think I see him shake his head and roll his eyes.

'It's part of the cleansing.'

In silence he removes the beautiful Incan tapestry that covered the Temple floor and is now a giant taco.

'You are wide open,' he says, the purge is a relinquishing of the last elements – Chile Heatwave Doritos, Reese's Pieces, a dozen Spearmint Extra that I have a habit of swallowing – of resistance.

'Am I?'

I am clinically suggestible is what I am. He is reloading the crack pipe.

'The preliminary dose is for me to read you, get a feel for what kind of numbers you need for breakthrough.'

'I don't think my body can take any more breakthrough.'

'Where you're going there is no body.' Touché baby! That's fine then ... Wait, what do you mean ... ?

But the pipe is stoked, the torch is on its way.

'Empty your lungs, then breathe in, slowly as you can, count it in ... a "yes" on the way in, "thank you" on the way out ... all the way to fifteen and hold it.'

One two ... This is quite mild, like a deep meditation ... three ... His eyes are turning brighter blue ... four, five, six, yes ... Apart from that, everything is in the right place ... seven, eight ... I think I've developed tolerance, I still have my children ... yes, nine ... Out of nowhere the childhood memory of swallowing a live frog to impress some friends ... ten ... Karma ... yes ... It's still in there, alive ... eleven ...

It's looking at me through his bright blue eyes ... twelve
... It's saying something to me through him, but the words
can't reach my ears ... thirteen ... My legs are disappearing
... yes ... His billboard-size surgeon face, I can see down the
holes in the pores in his nose, down the back of his throat
into his toad-thorax as my torso disappears ... yes ... fourteen
... Lips, the only bit left of me, looking into the desert of his
... native ... gringo-hating ... eyes ... lips puckering as the
toad kisses me ... fif ... Luis ... yes ... fift ... *Santa madre,
todo es vanidad!* (I had never spoken Spanish before)—

'Keep going, I got you.'

'Yes?'

No. Apart from the disappearing feeling of the tips of my
lips tightly pursed around the crack pipe, I am not there.

'Thank you.'

I am swallowed, inside the ancient creature, deep inside
the earth, in dark, wet clay. Inside me is that other toad, the
one I swallowed, inside my head, snout pushed against my
pineal gland. Inside him, there is the thought of me, a toad-
killer, thinking of him. Now it's his turn to purge and I am
rocketed, falling like a severed elevator which is really upwards
towards the hospital roof, below the wailing of nurses, who
love me. Somewhere far below, there is a grown man of forty-
eight, lying on a cushion caked in vomit on drugs looking up
at Amara leaning over me speaking in a child's voice:

'Do you love me?' she asks.

In fact, I am having contractions.

'I ...'

My cervix is dilating.

'Will you miss me when I'm gone?'

'...'

'Tell me the truth.'

'...'

Or this is how I imagine it in the zeptosecond that passes
between 'fifteen' and the exogenous agent smashing through

the blood brain barrier, flooding the clefts of the 100 billion neurons that make the circuitry of the fronto-limbic system, fusing thought, sensation, perception, mood, imagination, and beliefs about the self. But really it's not that at all, it's not remotely like it. Nor was it like other first-person accounts I read in the days afterwards on 'Third Wave', 'Tryptamine Palace' and other psychonaut sites. At least it wasn't to begin with. The signature of these was – like any mystical encounter, like Romantic love – its inexpressibility, the authors thrilling at their own defeat:

WOW – the Bomb. What-ever!
... an abysmal plunge into an incomprehensible abyss
... really the most intense experience possible, you just have to believe me on this

Or they toss desperately overwrought word salads:

my closed eyes were seeing bodies of undulations emanating infini-
tesimal fibres of existence carried off by another quadrillion overly
anxious nano-sided carnivorous ants

Like a key symptom of the drug is variations on aphasia. What actually happened was the thing that existed in the space after my expectations, and before my memory got to work revising it, before reading the contaminating accounts online, before discussing it with Amara, before confessing to my NA sponsor, before crying to Patrick. A space that, like the Line of Control, like an uncertainty principle, exists only as an idea, an idea that can't be stable once it's considered, or like metaphor itself, that goes from the unknown to the known with almost nothing to show for it, like this:

A deafening, high-pitched tone, the one that once signi-fied the end of the day's TV, when it still stopped for the night. Over which, a sky turned black except for a burning blue slash (the strip light on the ceiling?), which crackles into rectangular life as a TV display of clichéd 1960s kaleidoscopica; fractals moving in circles, Necker cubes, Escher walkways, in a tinny, garish two-tone orange and blue. Ahhhh. The feeling of relief – that these drugs are a dated, unthreatening spectacle, somehow tied up with 1960s kitsch. Then the thought that – hang on – this isn't happening on TV – my brain is generating this, of its own accord, without stimuli, which means – hang on – that 1960s psychedelic imagery and cheap topical illusions are a biological fact, true for eternity, inherent in brain chemistry, and not some sad contingent historical event. I start to panic at the implications. But – hang on, calm down – look, at least I know what's going on, and with that the thought that this won't be so difficult to write about after all. Look: I'm doing it. As soon as that thought appears, crowned by its 'I', it explodes leaving nothing in its place momentarily, a panic-ridden nothing which needs a new 'I' to quickly fill it. It only makes things worse. I recall the spiritual literature and think that giving up might be the only way out. Lie back, I got you. But there is still the 'I' dreaming of lying back. I need to self-soothe. Out loud I say, 'It's OK, I'm going to be—' BANG ... debris ... There is nothing to say soothing things, nothing to soothe, nothing to soothe with. Therefore panic. Therefore 'I—' BANG ... And that becomes the pattern: 'I—' BANG ... smithereens ... 'I—' BANG ... 'I—' BANG ... 'There are no patterns here—' BANG ... 'It's still—' BANG ... 'my—' BANG ... 'voice—' BANG ... 'Nothing filling up with watery terror, like a sand-hole at Broadstairs—' BANG ... 'like ... you can't ... say ... It's ... like ... Anything ... It's ... What? ... If ... I ... I ...' Every attempt to start, interpret, explain – all of them variant species of some naked, ravenous will, some self – is detonated at its point of origin, leaving dread, defeat, death. The end of man. Even that was a limit, when none such existed. Nothing will tolerate imagining here. BANG, BANG, BANG ... Our

BANG thoughts BANG and feelings BANG have no BANG need BANG of BANG us BANG ...

Then:

Eyes flick open like a toad's. There is dense foliage blocking out the sky. The figure, a naked middle-aged man, is lying on the jungle floor. Like Davis. He takes a deep breath. He knows that this is the first breath taken by the species, and that that knowing counts as its first thought; third-person Human Consciousness begins. Another moment passes and this one is recognisably mine; first-person consciousness begins. The He/I bathing in the sound of bells and singing bowls. I can't breathe but that's OK by him. In the corner of my eye the figure of a bare-chested man in prana yoga pants with the head of a condor. I am in ICU and he is looking after me. He is as high as I am. In this way, awareness as brief vignettes, stitching together perspectives like a slow strobing, more dark than light. Like Bella. He lays a blanket on me. I will need to breathe soon. I am under warm water, a pulse all around me, spinning slowly on my axis, like an astronaut inside her – it's either Amara, or Adora or Esme or Bella, unless it's my real mother. Either way she is crying, wailing and crying. She loves me. The bell is going to ring. I am about to come. I cough, my first cough, spit out something. It is either placenta or avocado. It lands back on my face. It will stay there forever. Weeping, laughing, coming. The waters are receding. I am breathing without the ventilator. Every time the voice which I recognise as mine thinks in terms of 'I' a bell sounds somewhere inside me, like an underwater chapel, to let me know the whole 'I' thing is just a joke. At last: the final door, no memory like Jim, but also no desire, no expectation, no understanding. My work here is done. I let my face go slack, dead, to tell its own truth without my input for the first time in its life. My legs float up towards my chest, naturally, allowing me to give birth more easily.

Bell by bell I watch myself return, then disappear, then return a little further, layer by layer. I accept it, I recognise every scintilla that goes into my reconstruction; the same as before and yet a slightly different copy of the original: fresh, un-fixated, unbiased, the real A. K. Benjamin. Lewis is behind me, holding my hand I think, and he is saying 'Happy birthday.' I try to speak but I have no language yet. My eyes open and she is where the sky was – Adora-Esme-Amara-Mother – the beat at the start of my universe – my head nestling in her lap, her upside-down face looking into mine, who is less than a minute old, already recognising her, already about to talk crap. She is smiling and weeping, like she has given birth.

'Just lie back, I got you,' she says.

'Fuck ... me ... Mama.' My first words.

Over the next hour more language returned, following the stages of childhood: pointing, babbling, words then short, telegram phrases. Or rather I can hear the delinquent drug-addicted version of me speaking, clambering for attention while the real me nests deeper inside, happy to stay there. Until I am telling Lewis, my new bromigo, that he's the real doctor here, that it must be so intoxicating, the power he has to annihilate, delight, transform. Then I tell him about how once during a brief foray into chromatic music I listened to Schoenberg, Webern and eventually Stockhausen, but I never relished any of it, not, that is, until I heard Alban Berg who in works such as the Violin Concerto and Kammerkonzert seemed to find the perfect balance between two poles; modernist enough to be strange, startling, and yet Romantic in the lushness of some of the progressions and cadences, and holding both together, at the same time, in a way that was so transporting, so exulting in its effect that I could actually feel my feet leaving the ground:

'Can you see what I'm saying Lewis ... Can you?'

I had a picture in my mind of music that would transform hungry, complex, fragile men into blue sexless angels. It turned out the picture already existed:

Kazuya Akimoto, *Blue Angel, before leaving (Homage to Alban Berg, Violin Concerto)*

'... You see that don't you ... ?'

Of course Lewis couldn't see it. He looked confused.

'Well that's what the trip was like and that's also the desired effect of the book I've been trying to write.'

And while I spoke there was still another version of me deeper inside, wondering why I felt the need to say such things, especially given that Lewis wasn't there. Next door I can hear Amara meeting a Skype commitment to attend a breath tutorial. I eat crisps, chew gum, turn my phone on, scratch my ears, catch myself sighing; the dull restless panic of being myself restored. But I'm not myself. Things are different.

Over the next few weeks we don't stray far from a cabin on the far corner of the compound. We switch our phones off and write automated-response emails for work. Like toddlers we eat, nap, take short, gentle walks away from base and then back again, but toddlers who only have each other.

I look at Amara and Amara looks at me. Looking at someone and keeping them real is like learning to breathe underwater. I am not used to this.

'Do me a favour?' she says.

'What's that?'

'When you write about this, leave me as I am, what I say, my hair, the spot on my nose, everything.'

'What spot?'

'Everything.'

'Of course, darling.'

I was lying. I kept all the original names in this book until the very last moment. It hurt so much, changing them, losing the people behind them, watching them disappear under the surface of something that staked its life on wanting to be real, like coins thrown into a wishing well when it's the only money you have left to get home. And when I changed them, I did it recklessly – chose bad names, joke names, the wrong names, which defaced their beautiful subjects, marred the truth of who they were. In fact the whole thing hurt so much I had to keep the odd one.

'After what we've just been through, I should be allowed to speak for myself.'

'You should. You will.'

'Together we're making a map of the world, as it is. I want to plant a flag right here, right now and say, "This is reality."' I'd never seen her look so intense. She's truly amazing this woman.

The hairs on the back of my neck bristle as I say, 'You are so nearly right my darling R*****, except the flag you speak of, it's over here now; this is reality.' She rolled her eyes. 'And this is the only moment we will ever have that stands any chance of being real at all.'

We leave the next morning, before dawn can bend itself over the top of the hills, driving down through ponderosa pines towards the ocean, then heading north back towards the

Bay. It's early April, no cars to speak of, before the tourists start in earnest. Turning a blind corner, there's a wild dog standing in the middle of the road waiting for us with something in its mouth. 'Coyote. They never normally come this far down from the mountains,' Amara says. We are not supposed to be seeing this.

There were still no cars when we hit the Pacific Coast Highway. The ocean looked more still than I remembered it. Things really were different, it was inescapable; just as I was ready to declare that 'the most powerful psychedelic known to man' was faux-scientific hippy marketing. Everything looked more still, more quiet, more real. There was a solitary walker on the pavement wearing a face mask. And then no one for hundreds of miles.

Epilogue

They all brought me here, the future, and the understanding that not everything I touch need turn to memoir, that I am free to divorce from my own antic creations, beginning with A. K. Benjamin, old enough to know that the propositions of the wind in the long grass or the knowledge that everyone you have ever met will one day lose everyone they have ever cared about, will forever remain beyond my grasp. It's time to tell someone else's story – courage, honour, service, love, madness, murder – without getting in the way, let the wildness that stands in no need of our fabricating or consummation, that is always ecstatically there – brilliant as violet unblinking eyes – shine through.

But first my promise to her.

I think about the moment, late in the afternoon, and the type of restlessness that would often set in for her at that hour, when the day's idiom feels decided, its current caught. I imagine it's the determination to change course that brings her out onto the patio, really a mosaic of local chalk stone, to the lip of the pool. The thought of starting again. The afternoon air is thick and sweet, like the espresso she drank twenty minutes before. Though the stone is painfully hot underfoot, she is perfectly still, when 'still' was still a choice. Before her the water is as still as she is, its skin dimpling under floating leaves, small black insects dancing on their backs. Behind her the sun is a thin white smear, lying like an Indian ascetic on the tips of knife-like alps. The mountains are centaurs here, forested to the waist, naked black rock above. Below them, reflections from the house's windows catch fire in Lake Shkodra, bordered by dying palms where starlings and parakeets take turns to teem. But even they are still in this moment; water, animals, air, entranced by the same spell, as though her body could extend its will to everything.

But it's not power she's feeling, it's paralysis. Because once stillness has been thought about, and one reflects on what has to happen for things to move at all, then even the slightest gesture becomes difficult. Stillness wins, its own glue. The moment stretches out further. The longer she doesn't move the more she feels outfaced by the reality of water – one of the infinite ways the world pushes back. She remembers the childish hours she devoted to willing time, the elements, other people to do her bidding. Sixty years later, still the same magical thinking. How ridiculous that she should have a pool.

Really her stillness is a mirage, a surface under which an unimaginable cascade of activities are under way. I have the fantasy that somewhere, amidst the sensations of standing up, the heat of the patio, the general trembling underneath still-ness, the wall of thick sweet air landing on two square metres of skin, the droplet of sweat like a lover's finger slaloming down the inside of her arm – that amidst this cloud of sensa-tions there is the tiny ghost of the feeling of the blood building against the dam in her artery that will burst terribly a moment from now. Why have we evolved to be so insensitive? Where is the pain we need?

Her white bathing suit, printed with monstrous-looking carnations, pinches a little. There hadn't been time for tennis this year, wrapping things up at work, busy making sure her successor had the easiest possible way of grasping how complicated her job was. She finds her gaze has fallen on the space between her and the pool's surface, on pure hovering air. She's never seen it like this before, rather than compressing the world into pictures, missing the depth of what's really there. The sunlight has just got brighter, blitzing the water, so that the contrast between pool and the white stone patio has gone, and with it any proof that the two surfaces are different: a toddler might step out from the patio's edge without hesitation, disappearing as noiselessly as a toad from a rock. No grandchildren and therefore no need to put a fence with a gate round the water. No children because, as she would say, she didn't want to redeem her own childhood through them. It might have been different if she'd met Marc a decade earlier, but it probably wouldn't. Really, not having children, like this stillness, like many of the states that describe her, was neither decided nor accidental.

Childless. It still felt shocking as a word, new facets of it yet to be taken in, old ones reconsidered. Like how she could sometimes identify with the child she didn't have, see herself as a ghost; because she lacked substance, because something remote in her always sought out rejected things, because her long-dead father still controlled the currents by which her life secretly moved. Or didn't move. The sky was silver foil for a moment – the smear of sun thin and neat as a line of heroin – making the air feel that much hotter, like she was the thing being baked. She has all the time in the world, that's what people kept saying. It wasn't true, she had no time at all; things were slashed and burnt as soon as she'd experienced them, the last day of work already only as bright as the first, the days ahead merely a few dull items of more or less familiar furniture stacked in the foreground, giving way to an empty house. What lay ahead, meanwhile, had become strangely viscous; to break through its membrane would require something more than her will. She had come out onto the patio determined to dive in and change the course of the afternoon, but now she doesn't want to wound the water.

The house was a modest bungalow near the town square, a dilapidated cafe on one side, a young local family on the other. It had been the mayor's before he died. Near the end of a long illness he made a gift – absurd or astute depending on who you asked – to every residence: an English rose with a trellis to help it climb the house. To open hearts. To woo tourists. Except now the plants looked like desiccated snakes in the June heat. The pool apart, the house was modest and sensible, in good repair, on one floor, no garden to maintain. It was the sort of place one might grow old in or, as sixty-two was already old, very old in. The signal to dive was on its way. There were risks, of course. Even back then, several years ago, she had quantified the possibility of the UK leaving Europe and how that might change things for them. Medical complications and their insurance were foremost in her mind. Statistically it was highly improbable she would suffer a so-called serious condition for another eight to ten years, and given her health background and socio-economic status it was more like twelve to fifteen. Maybe she hallucinated the signal being sent, maybe part of her prefers to be stuck, a training for the time that comes to most when she won't be able to do what she wants, even the most basic things. The soft silver light is beginning to fade, deepening the blue of the water, an ongoing invitation. The first dull stars are low in the sky.

What mattered was not catastrophe and its audit; that had been her entire working life. What mattered was the thought of those mountains nearby, bordering one of the last remaining wildernesses of any significance in mainland Europe, occasionally broached in search of a missing person or to trade a mule but wholly 'un-hiked'. There was only one existing map, made by the Russians over fifty years ago, which was low-scale and quite possibly, for strategic reasons, deliberately unreliable. To have something near to her that remained permanently strange, it felt like a sign of health. There was no mobile-phone reception out there. Angola had reception, East Timor had reception, but not out there. The fact that it hadn't been covered, satellited, 5G'd, Facebooked, bucket-listed, was profoundly attractive to her after half a century living in a place where everything, including the most private thoughts in her head, had become progressively more known. Actual conversation would also be restricted given there was nobody who spoke anything more than extremely broken English in the town. And she had decided – that word again – not to learn the language, not a single word of it, before she moved here. So she wouldn't have known that 'stroke', so feminine, so ironic in English, becomes the equally tender *'pash'* in Albanian. These preferences, taken together, were uncharacteristic; signs of incipient wildness. A dementia on the way? It was possible, but improbable. Signs of a long, deep pining for change, for something uncharacteristically sudden, for total immersion; the unconscious shunning of order and safety in favour of a territory with a history of war, massacre, natural disaster, because she finally wants to experience the parts of herself that have always been in conflict.

Her dive seemed flatter than normal, a racing dive, cracking the water like a starter's gun, enough to send distant parakeets fizzing into the air. Marc watches her from behind his easel in the living room, already a little swamped by his retirement commitment to teach himself oil painting. It's still there, the thrill at her easy athleticism; the type of body that will harbour youthfulness right to its end, he thinks. Underwater, the carnations seem to run into one another, like she's been shot in the back. Such a beautiful swimmer; with her reach she would need only one or two full strokes to make it to the other end without surfacing. He imagines the silence of being underwater. Real retirement was being free of one's voice, of any attribute that anchored you – another reason to move to the southern Balkans – so that it wasn't retirement at all, but a weightlessness like swimming, a water birth that would begin at the moment her fingertips touch the pool's surface, which was now settling down again, leaves and dead insects bobbing towards the grill at the edge, while her body turns gently over and glides to rest on the pool's bottom, never to move again by its own accord.

Lie back, lie back on the inside too; accept your new element; it's all we can do when reality presses down, I want to tell her.

There, in the fraction of a moment before she disappears completely (Marc has seconds to tear out of the house, dive into the pool, and bring her youthful body up to the surface) – the moment on which everything hinges – and after which the two of us, Bella and I, must part ways – she is imagining that she is heading toward a future in which her days will be much the same as they always have been, changing of course, but imperceptibly. We might hope that her very last thought while she is still on this side of the door would be of Marc, her love for him, proof that something genuinely new was still a possibility, that not all things decline. That morning she had been walking behind him up a steep, rocky path, the town a long way below them, even the terraces of farmland had flattened out from this height. They must have been at 2,000 metres, only one or two trees above them. She could feel the blood pumping freely from her heart through her neck, down through her legs, moving through her feet. Her stride got longer, exulting in the feeling of lifting off, of heading into the clouds. In front of her he shuffled with much smaller steps, legs bowed after years of running, his bald spot bobbing up and down like a tiny moon before her. She had been longing to lose herself in these alps for years. At last she was here, with the man who had waited for her to find him. It was unexpected freedom, as though finally after years of impossible struggles and dead ends she had found someone with whom she could breathe underwater. She felt this overwhelming urge to bound up to him, hook her arms under his and squeeze him, as though she could make the blood circulating in each of them one and the same. At least she wanted to stop him in his tracks, tell him what a miracle he had made possible by seeing things through with her. But that could wait a moment. Just watching him as he was, approaching sixty, bowed, stoic, gentle, solid, unwatched – carrying a pack for both of them with lunch, extra jumpers, waterproofs – made her memory burn with the times they had shared these last ten years, love's roll call: the moment when he first pressed her hand outside the Lawnswood Arms; how she had been so shy she didn't call him back for a month after they kissed; how angry he was with her for that, then how soft; and the time he sang to her while she read her favourite poetry in a lavender-scented bath, then looking up through dark, bright eyes filled with every kind of delight in time to hear his voice break through to a deeper, more fragile, more beautiful version of itself, which took her breath away; his finally being able to sing unselfconsciously, as though she wasn't there.